Kinematics - Analysis and Applications

Edited by Joseph Mizrahi

Published in London, United Kingdom

IntechOpen

Supporting open minds since 2005

Kinematics - Analysis and Applications
http://dx.doi.org/10.5772/intechopen.77610
Edited by Joseph Mizrahi

Contributors
Ijar M Da Fonseca, Maurício N. Pontuschka, Glaydson Luiz Bertoze Lima, Stanislav Barton, Baokun Li, Guangbo Hao, Jiří Ondrášek, Sebastian Linß, Stefan Henning, Lena Zentner, Changsheng Cai, Lin Pan, Jianjun Zhu, Xianqiang Cui

First published in London, United Kingdom, 2019 by IntechOpen
IntechOpen is the global imprint of INTECHOPEN LIMITED, registered in England and Wales, registration number: 11086078, The Shard, 25th floor, 32 London Bridge Street
London, SE19SG – United Kingdom
Printed in Croatia

British Library Cataloguing-in-Publication Data
A catalogue record for this book is available from the British Library

Additional hard and PDF copies can be obtained from orders@intechopen.com

Kinematics - Analysis and Applications
Edited by Joseph Mizrahi
p. cm.
Print ISBN 978-1-78984-490-0
Online ISBN 978-1-78984-491-7
eBook (PDF) ISBN 978-1-78985-700-9

For EU product safety concerns:
IN TECH d.o.o., Prolaz Marije Krucifikse Kozulić 3, 51000 Rijeka, Croatia, info@intechopen.com or visit our website at intechopen.com.

Meet the editor

Professor Emeritus Joseph Mizrahi is a faculty member of the Biomedical Engineering (BME) Department at Technion – Israel Institute of Technology. He received his BSc in Aerospace Engineering; MSc in Mechanics; and PhD in Biomedical Engineering, all from Technion. He was Chair-Professor and served as Head of the BME Department for five years. He has also held positions with the Universities of the Witwatersrand Johannesburg, Cape Town, Harvard, Hong Kong Polytechnic, Drexel, and NCKU in Taiwan. He has headed for 18 years the Biomechanics Laboratory at the Loewenstein Rehabilitation Center in Israel. His research interests, in Orthopaedic Biomechanics and Rehabilitation Neuro-Engineering, include: musculo-skeletal mechanics; muscle/bone interactions; muscle fatigue; Functional Electrical Stimulation of excitable tissues; and tissue engineering, musculo-skeletal redundancies, and mechanical indeterminacies. He has authored more than 250 publications, including journal papers, conference proceedings, book chapters, and two books.

Contents

Preface

Kinematics deals with changes of position (motion, displacement, velocity, acceleration, etc.) and/or geometry (strain, strain rate, volume, etc.), without being concerned with the causes of these changes (forces, torques, etc.). Numerous problems in engineering and biology can be described, formulated, characterized, and analyzed in kinematics terms.

In classical machinery and robotics the most distinctive characteristic is constrained motion of interconnected parts relative to one another. Thus, multi-degree-of-freedom kinematic chains, robotic arms, and manipulators have become essential devices in industrial applications, including manufacturing, vehicle, space, agriculture, military, and medicine. Impressive recent developments in medical robotics have included surgical robots and robots for aiding persons with disabilities. Miniaturization has led to nano-manipulators and robots. Kinematics is also used as a basis for designing sensors such as displacement transducers, accelerometers, gyroscopes, and strain gauges.

In biological organs kinematics serves to describe motion, deformation, or both. Multi-segment representation of the human body, with the segments usually treated as rigid bodies, enables to describe sets of movements, including standing, walking, running, and jumping. More recent analyses have extended the multi-segment representation to express effects of wobbling mass due to influence of the soft internal organs in the limbs, abdomen, chest, skull, and the like. Kinematics of deformable bodies, such as the beating heart or uterus under labor contractions, can be expressed in terms of strain, geometric curvature, and local obliqueness, providing essential information about the deforming organ. At the micro level, kinematics is being applied in investigations of cellular response (motion and deformation) through mechanical changes induced to the medium in which the cells are being cultured.

This book provides the reader with an updated look at the current trends in kinematics methods and applications. It comprises of three sections. The first section deals with kinematics of linkages and includes analysis of cam mechanisms, kinematic pairs, and transformation of rotary motion into oscillation. The second section covers compliant mechanisms, whereby elastically deformable parts rather than conventional joints are part of the mechanism. The third section deals with kinematics of spacecrafts and satellites in the contexts of global navigation systems as well as space robot analysis using forward and inverse kinematics.

Joseph Mizrahi
Professor
Department of Biomedical Engineering Technion,
Israel Institute of Technology,
Haifa, Israel

Section 1

Kinematics of Linkages

Chapter 1

Optimization of Kinematics of Inclined Swinging Pin

Stanislav Barton

Abstract

The inclined swinging pin simply transforms the rotary motion into a rotational oscillation. It consists of three components, a rotating inclined tenon, a crossbeam, and a stirrup. The pitch angle of the inclined tenon relative to the x-axis of its rotation is decisive for the kinematics of this joint. Too small inclination angle will result in small amplitude of oscillation; too much inclination will lead to an impulse dynamic stress that can damage the pin. The optimal angle of inclination can be determined from the mathematical model, created in the Maple environment, which uses linear algebra resources to describe the behavior of the joint. Vectors of coordinates of the critical points are created for each joint component. Furthermore, transformation matrices are created which describe the behavior of the corresponding kinematic chain in relation to the inclination angle of the inclined tenon and time.

Keywords: mathematical modeling in Maple, transformation matrix, angular velocity, angular acceleration, optimization

1. Introduction

The inclined swinging pin is used for the simple conversion of rotational motion into oscillation in many agricultural machines. For example, in combine harvesters, it is used to drive a mowing bar that cuts stems. It is composed of three basic components: an inclined tenon—red, a crossbeam—green, and a stirrup—blue, see **Figure 1**. We study the movement of individual components in a rectangular coordinate system, see **Figure 1**.

Similar mechanisms are described, for example, in the book [1] that is available in an electronic version [2]. Many animations can be found on the Web, if you use keywords *rotation to oscillation mechanism*, Google will offer you over 20,000 links.

2. The basic components

2.1 Tenon

The pin is a basic component. Its basic part is an oblique nose. Its axis of symmetry forms an angle ϕ with the axis of rotation of the whole pin, which is identical to the axis x of the coordinate system. The axis of symmetry of the nose

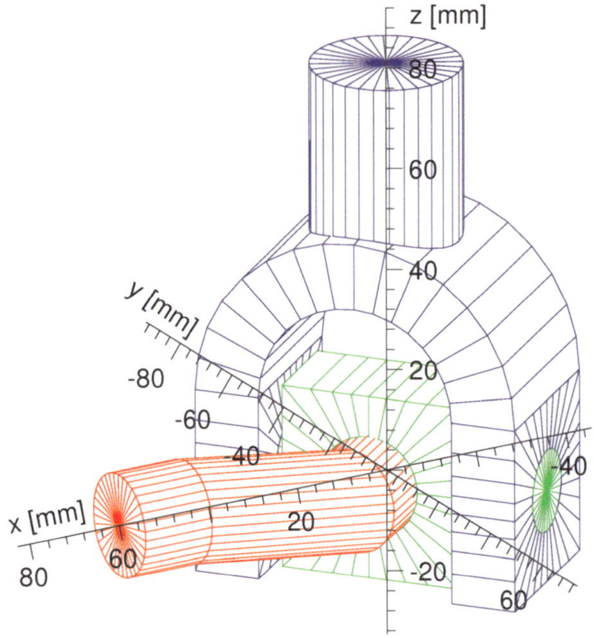

Figure 1.
Joint-components and coordinate system.

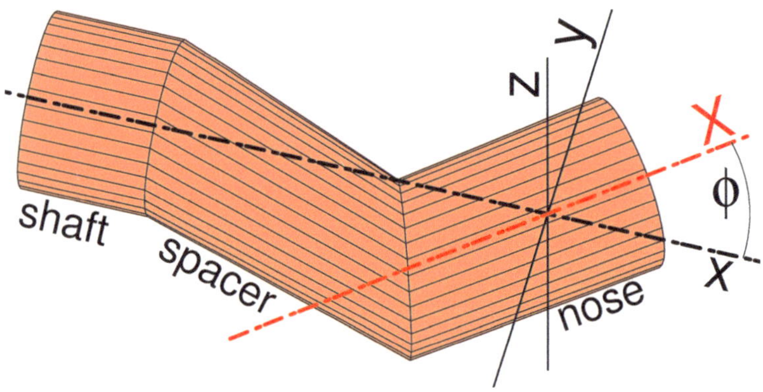

Figure 2.
Tenon.

passes through the beginning of the coordinate system. The spacer serves to secure the nose to the support shaft, see **Figure 2**.

2.2 Crossbeam and stirrup

The crossbeam is the middle member of the kinematic chain. The tenon nose is pushed through its center hole. As the nose rotates, the crossbar turns. The stirrup is attached to its side pins, and therefore the axis of the side pins is constantly in the plane $z = 0$, see **Figure 3**.

Depending on how the crossbeam swings, the stirrup oscillates. The driven device connects to the vertical axis of the stirrup, see **Figure 4**.

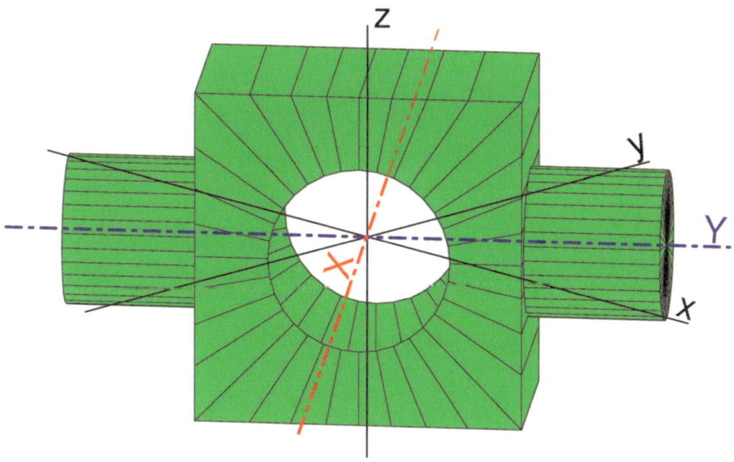

Figure 3.
Crossbeam.

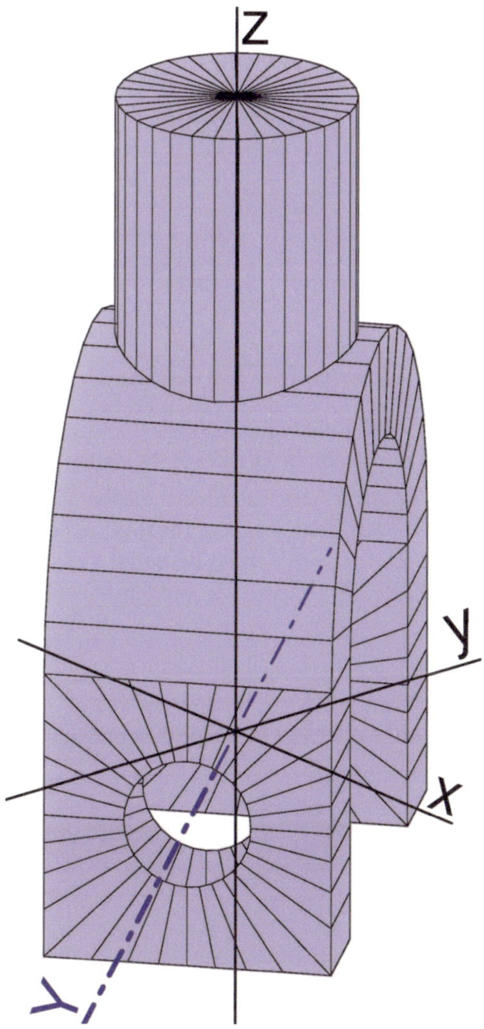

Figure 4.
Stirrup.

3. Deriving of transform matrices

We will derive the transformation matrices in the Maple13 program environment. The decisive element for which the transformation matrix will be derived is the crossbeam.

Allow us to specify axes of the base coordinate system in which tenon—**Figures 1** and **2**—are displayed as $[x, y, z]$. The axes of the moving coordinate system in which we plot the actual position of the crossbeam are denoted as $[X, Y, Z]$. Let us denote the transformation matrix from system $[x, y, z]$ into system $[X, Y, Z]$ as *R_xyz*. The individual rows of the *R_xyz* matrix are the x, y, z coordinates of the unit vectors in the X, Y, Z direction.

The direction of the X-axis is entered as the direction of the tenon's nose axis. The axis X, which is equal to the axis of the nose, creates the angle ϕ with the x-axis and generally the entire tenon is rotated by an angle ψ around the x-axis.

A detailed description of the properties of transformation matrices can be found, for example, in [3].

```
> restart; with(LinearAlgebra): with(plots): #Maple Start+Libraries
> R_z:=Matrix([[cos(phi),sin(phi),0],
   [sin(phi),cos(phi),0],[0,0,1]]): #R_z - Rotation around z
> R_x:=Matrix([[1,0,0],[0,cos(psi),sin(psi)],
   [0,-sin(psi),cos(psi)]]):]]): #R_x - Rotation around x
> R_xz:=R_x.R_z;
```

$$R\_xz := \begin{bmatrix} \cos(\varphi) & \sin(\varphi) & 0 \\ -\cos(\psi)\sin(\varphi) & \cos(\psi)\cos(\varphi) & \sin(\psi) \\ \sin(\psi)\sin(\varphi) & -\sin(\psi)\cos(\varphi) & \cos(\psi) \end{bmatrix} \tag{1}$$

```
> X:=convert(R_xz.<1,0,0>,list);
```

$$X := [\cos(\varphi), -\cos(\psi)\sin(\varphi), \sin(\psi)\sin(\varphi)]. \tag{2}$$

Equation (2) specifies the direction of the nose axis, thus X-axis vector in the coordinates $[x, y, z]$. Coordinates Y are calculated easily. We know that $|Y| = 1$ and Yz must always be equal to 0 and that X and Y must be perpendicular to each other, i.e., $X{\cdot}Y = 0$, the scalar product X and Y must be equal to 0. Therefore, we choose the vector coordinates $Y = [Yx, Yy, 0]$ and Yx and Yy are calculated from the following equations:

```
> Y:=[Yx,Yy,0];
> E1:=add(w,w=zip((u,v)->u*v,X,Y)); E2:=Yx^2+Yy^2=1;
> simplify(allvalues(solve({E1,E2},{Yx,Yy}))[2],symbolic);
> Y:=subs(%,Y);
```

$$Y := \left[\frac{\cos(\psi)\sin(\varphi)}{\sqrt{\cos(\psi)^2 - \cos(\psi)^2\cos(\varphi)^2 + \cos(\phi)^2}}, \frac{\cos(\varphi)}{\sqrt{\cos(\psi)^2 - \cos(\psi)^2\cos(\varphi)^2 + \cos(\phi)^2}}, 0\right]. \tag{3}$$

Equations **E1** and **E2** have two solutions. In order for the XYZ system to be oriented counterclockwise, a second solution has to be chosen. Coordinates Z are calculated from the conditions $X{\cdot}Z = 0$, $Y{\cdot}Z = 0$, and $|Z| = 1$.

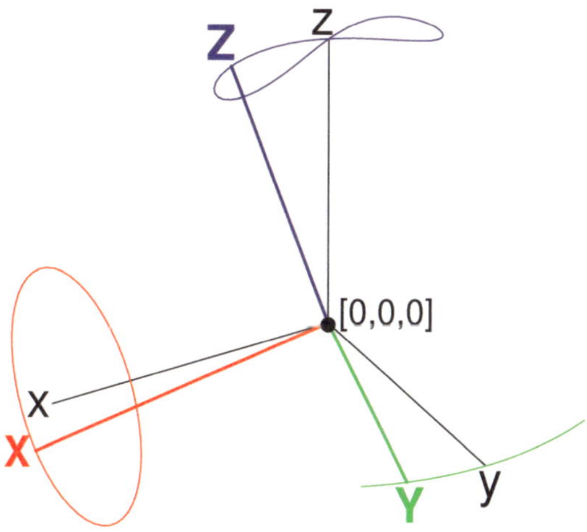

Figure 5.
Movement of the unit vectors X, Y, Z.

```
> Z:=[Zx,Zy,Zz];
> E1:=add(w,w=zip((u,v)->u*v,X,Z));
> E2:=add(w,w=zip((u,v)->u*v,Z,Y)); E3:=Zx^2+Zy^2+Zz^2=1;
> simplify(allvalues(solve({E1,E2,E3},{Zx,Zy,Zz}))[1],symbolic);
> Z:=subs(%,Z);
```

$$Z := \left[\frac{\sin(\psi)\cos(\varphi)\sin(\varphi)}{\%1}, \frac{\sin(\psi)\cos(\psi)\sin(\varphi)^2}{\%1}, \%1\right], \quad \text{where } \%1 = \sqrt{\cos(\psi)^2 + \cos(\psi)^2\cos(\varphi)^2 + \cos(\varphi)^2}. \tag{4}$$

From the vectors X, Y, Z, it is possible to create a transformation matrix M_XYZ, which will describe the orientation of the crossbeam depending on the angle ϕ and the angle of rotation ψ.

```
> M_XYZ:=Matrix([X,Y,Z]);
```

$$M\_XYZ := \begin{bmatrix} \cos(\varphi) & -\cos(\psi)\sin(\varphi) & \sin(\psi)\sin(\varphi) \\ \dfrac{\cos(\psi)\sin(\varphi)}{\%1} & \dfrac{\cos(\varphi)}{\%1} & 0 \\ -\dfrac{\sin(\psi)\sin(\varphi)\cos(\varphi)}{\%1} & \dfrac{\sin(\psi)\cos(\psi)\sin(\varphi)^2}{\%1} & \%1 \end{bmatrix}. \tag{5}$$

Finally, it is possible to create a graph showing the curves described by the endpoints of the X, Y, and Z vectors. For instance, if $\phi = 2\pi/15$ and $\psi \in \langle 0, 2\pi \rangle$ are selected, then the resulting graph is **Figure 5**. The corresponding animation is stored in the file An_Axes.avi, available to download in the Videos section.

4. Optimization

The orientation of the Y axis can be derived from the matrix M_XYZ, see Eq. (5). The orientation of this axis indicates the angle of rotation of the stirrup about the z axis.

```
> Y:=evalm([0,1,0].M_XYZ);
```

$$Y := \left[-\frac{\cos(\psi)\sin(\varphi)}{\%1}, -\frac{\cos(\varphi)}{\%1}, 0\right]. \tag{6}$$

The angle of rotation of the stirrup is equal to the angle between the y and Y axes. Let us denote this angle as θ.

```
> theta:=arctan(rz[1]/rz[2]);
```

$$\theta := \arctan\left(\frac{\cos(\psi)\sin(\varphi)}{\cos(\varphi)}\right). \tag{7}$$

The angle ψ indicates the angle of rotation of the pin about the x-axis. We can therefore substitute:

```
> su:=psi=2*Pi*tau:
```

$$su := \psi = 2\pi\tau. \tag{8}$$

```
> theta:=subs(su,sin(phi)=cos(phi)*tan(phi),theta);
```

$$\theta := \arctan(\cos(2\pi\tau)\tan(\varphi)), \tag{9}$$

where $\tau = t/T$ = dimensionless time, t = time, and T = duration of one revolution = one period.

It is now possible to determine the angular velocity of the stirrup oscillations, $\omega = d\theta(\tau)/d\tau$, and the angular acceleration of the stirrup oscillations, $\varepsilon = d^2\theta(\tau)/d\tau^2$.

```
> omega:=diff(theta,tau);
```

$$\omega := -\frac{2\sin(2\pi\tau)\pi\tan(\varphi)}{T\left(1+\cos(2\pi\tau)^2\tan(\varphi)^2\right)} \tag{10}$$

```
> epsilon:=normal(diff(omega,tau));
```

$$\varepsilon := -\frac{4\cos(2\pi\tau)\pi^2\tan(\varphi)\left(1+\cos(2\pi t)^2\tan(\varphi)^2+2\sin(2\pi t)^2\tan(\varphi)^2\right)}{\left(1+\cos(2\pi t)^2\tan(\varphi)^2\right)^2} \tag{11}$$

Now we plot the graph of $\varepsilon(\tau)$ for $\tau \in \langle 0, 1\rangle$. The values of ϕ will be chosen between 5 and 45° with a step of 5°, see **Figure 6**.

```
> EP:=plot([seq(epsilon,phi=expand(Pi/36*[$1..9]))],
tau=0..1,color=red): EP;
```

It is quite clear from **Figure 6** that with an increase in angle ϕ—that is, the angle between the pin nose and the x-axis—the course of the angular acceleration of the oscillations begins to deviate sharply from the sinusoidal shape. Thus, large variations in angular acceleration will occur for large ϕ values, resulting in rapid wear of the entire mechanism. This can lead to mechanical damage, because high angular acceleration values require high torque forces. Conversely, small values of the angle ϕ will lead only to a low oscillation amplitude, which may not be sufficient for the proper operation of the driven device. It is therefore necessary to find the optimum angle value ϕ.

After studying the graphs in **Figure 6**, the value at which the angular acceleration for τ = 1/2 reaches the local maximum possible value appears optimal. If the

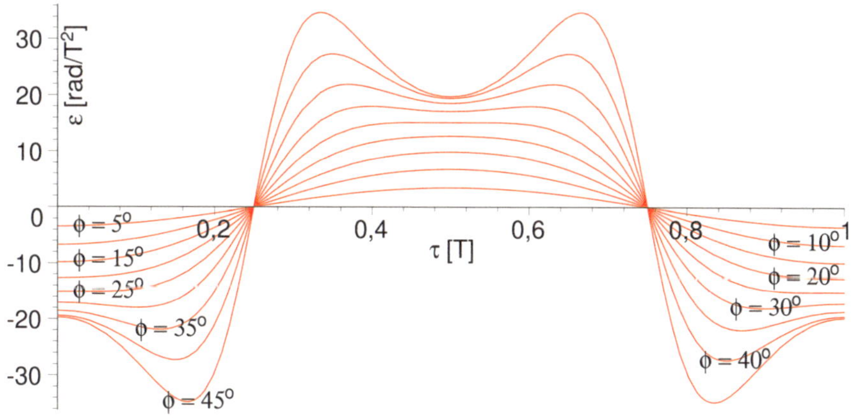

Figure 6.
Angular acceleration for ϕ = 5, 10, 15, 20, 25, 30, 35, 40, and 45°.

value of ϕ further increases, then this local maximum becomes the local minimum. This means that the optimum value of the angle ϕ must be the one at which the transition from local maximum to local minimum occurs. From a mathematical point of view, the following condition must be satisfied:

$$\left.\frac{d^2\,\varepsilon(\tau,\varphi)}{d\tau^2}\right|_{\tau=\frac{1}{2}} = 0. \tag{12}$$

The value of the angle ϕ can be easily determined from Eq. (12) using Maple.

```
> Ett:=normal(diff(epsilon,tau,tau));
```

$$\begin{aligned} Ett := &-16\cos(2\pi\tau)\pi^4\tan(\varphi)\,(-1+24\sin(2\pi\tau)^4\tan(\varphi)^6\cos(2\pi\tau)^2 \\ &+8\sin(2\pi\tau)^2\tan(\varphi)^4\cos(2\pi\tau)^2-20\sin(2\pi\tau)^2\tan(\varphi)^2+9\cos(2\pi\tau)^4\tan(\varphi)^4 \\ &+5\cos(2\pi\tau)^6\tan(\varphi)^6-24\sin(2\pi\tau)^4\tan(\varphi)^4+3\cos(2\pi\tau)^2\tan(\varphi)^2 \\ &+28\sin(2\pi\tau)^2\tan(\varphi)^6\cos(2\pi\tau)^4)/\left(1+\cos(2\pi\tau)^2\tan(\varphi)^2\right)^4 \end{aligned} \tag{13}$$

```
> Ett2:=eval(subs(tau=1/2,Ett));
```

$$Ett2 := \frac{16\,\pi^4\tan(\varphi)\left(-1+9\tan(\varphi)^4+5\tan(\varphi)^6+3\tan(\varphi)^2\right)}{\left(1+\tan(\varphi)^2\right)^4} \tag{14}$$

```
> Sol:=[solve(Ett2,phi)];
```

$$Sol := \left[\arctan\left(\frac{\sqrt{5}}{5}\right), -\arctan\left(\frac{\sqrt{5}}{5}\right), 0\right] \tag{15}$$

```
> Sold:=map(u->evalf(convert(u,degrees)),Sol);
```

$$Sold := [24.09484255\,\text{degrees},\quad -24.09484255\,\text{degrees}, 0\,\text{degrees}] \tag{16}$$

```
> Phi:=Sol[1];
```

$$\Phi := \arctan\left(\frac{\sqrt{5}}{5}\right). \tag{17}$$

```
> Epsilon:=simplify(subs(phi=Phi,epsilon),symbolic);
```

$$E' := \frac{4\pi^2\sqrt{5}\cos(2\pi\tau)\left(-7+\cos(2\pi\tau)^2\right)}{25+10\cos(2\pi\tau)^2+\cos(2\pi\tau)^4}. \quad (18)$$

It is now possible again to plot curves describing the angular acceleration dependence on τ and ϕ, and highlight the curve for $\phi = \Phi$, see **Figure 7**.

```
> EO:=plot(subs(phi=Phi,epsilon),tau=0..1,color=blue,thickness=3):
> display({EP,EO});
```

To make sure that the angle Φ is optimal, we will plot the angular velocity depending on τ, see **Figure 8**.

```
> Omega:=simplify(subs(phi=Phi,T=1,omega),symbolic);
```

$$\Omega := -\frac{2\pi\sqrt{5}\sin(2\pi\tau)}{5+\cos(2\pi\tau)^2}. \quad (19)$$

```
> OP:=plot([seq(omega,phi=expand(Pi/36*[$1..9]))],
        tau=0..1,color=red):
```

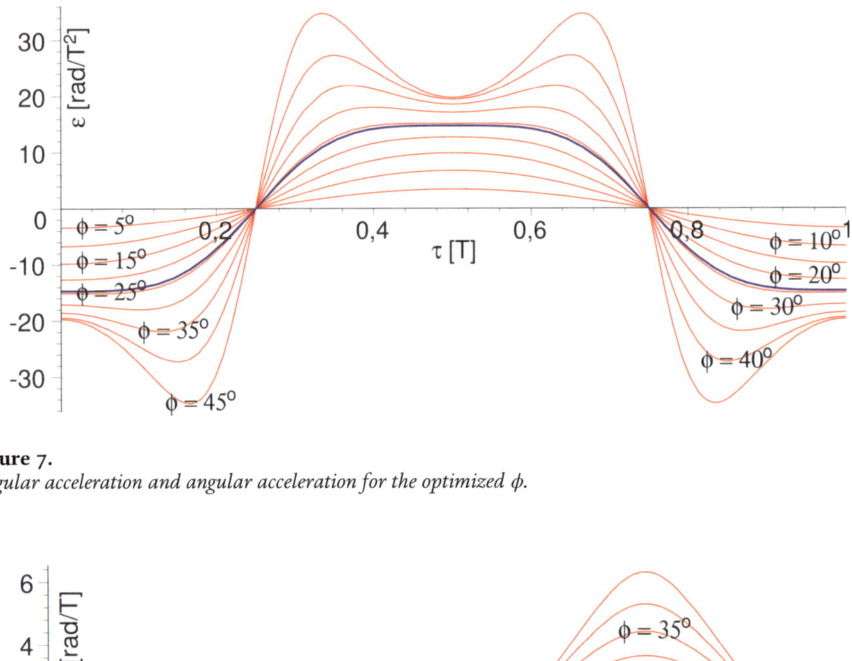

Figure 7.
Angular acceleration and angular acceleration for the optimized ϕ.

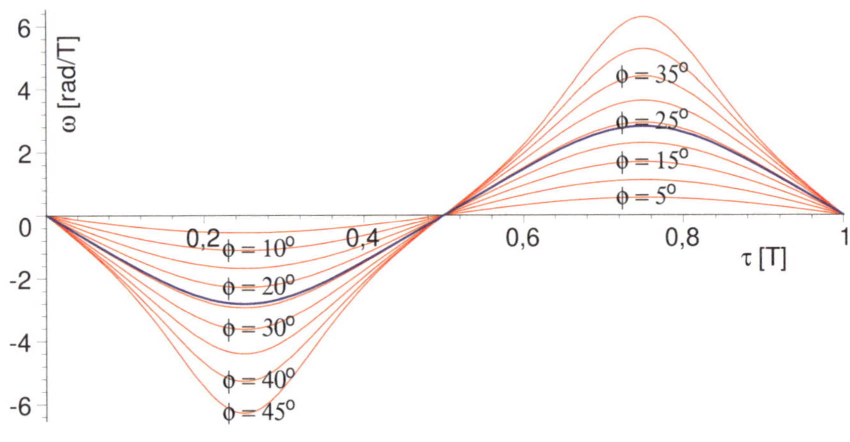

Figure 8.
The angular velocity of the oscillations.

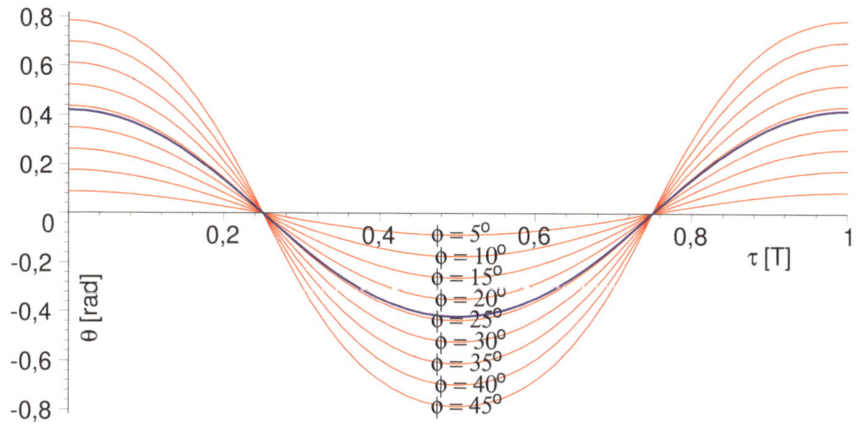

Figure 9.
Dependence of angle θ on τ.

```
> OO:=plot(Omega,tau=0..1,color=blue,thickness=3):
> display({OP,OO});
```

And yet it is possible to plot the dependence of the angle θ on τ, see **Figure 9**.

```
> Theta:=simplify(subs(phi=Phi,theta));
```

$$\Theta := \arctan\left(\frac{\sqrt{5}}{5}\cos(2\pi\tau)\right). \tag{20}$$

```
> TP:=plot([seq(theta,phi=expand(Pi/36*[$1..9]))],
      tau=0..1,color=red):
> TO:=plot(Theta,tau=0..1,color=blue,thickness=3):
> display({TP,TO});
```

5. Plotting of the dependencies in phase spaces

Further evidence of the choice of the optimum angle value Φ, that is, the choice of a suitable compromise between the oscillation amplitude and smooth oscillation over time, can be given by plotting the phase diagrams, which are graphs in which one kinematic variable is plotted against another kinematic variable.

In this case, we will present three phase diagrams, where we will plot the dependence of the oscillation angular velocity on the current value of the oscillation angle—$\omega(\theta(\tau))$, see **Figure 10**, then the course of the angular acceleration of the oscillations against the current value of the oscillation angle—$\varepsilon(\theta(\tau))$, see **Figure 11** and finally the course of the angular acceleration of the oscillations versus the angular velocity of the oscillations—$\varepsilon(\omega(\tau))$, see **Figure 12**.

```
> TW:=plot([seq([theta,omega,tau=0..1],
      phi=expand(Pi/36*[$1..9]))],color=red,numpoints=500):
> TWO:=plot([Theta,Omega,tau=0..1],color=blue,thickness=3):
> display({TW,TWO});
> TE:=plot([seq([theta,epsilon,tau=0..1],
      phi=expand(Pi/36*[$1..9]))],color=red,numpoints=500):
> TEO:=plot([Theta,Epsilon,tau=0..1],color=blue,thickness=3):
> display({TE,TEO});
> OE:=plot([seq([omega,epsilon,tau=0..1],
```

```
    phi=expand(Pi/36*[$1..9]))],color=red,numpoints=500):
> OEO:=plot([Omega,Epsilon,tau=0..1],color=blue,thickness=3):
> display({OE,OEO});
```

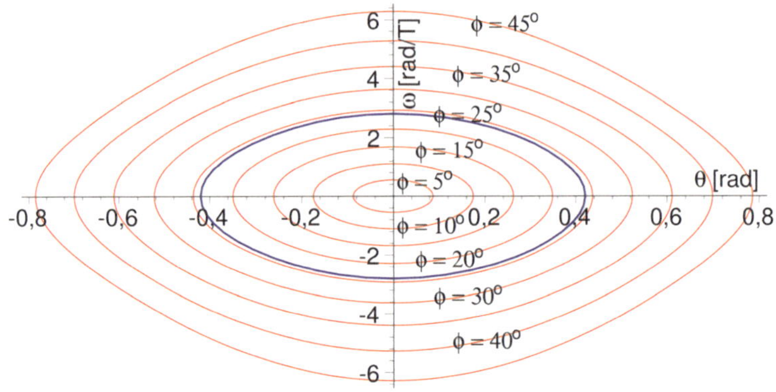

Figure 10.
The phase diagram $\omega(\theta(\tau))$.

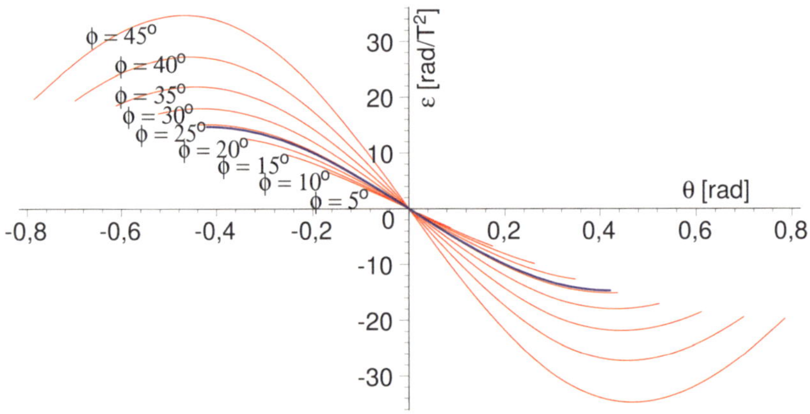

Figure 11.
The phase diagram of $\varepsilon(\theta(\tau))$.

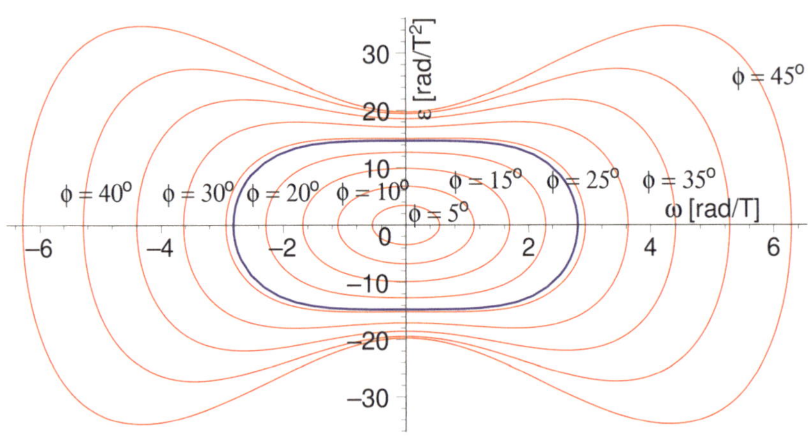

Figure 12.
The phase diagram of $\varepsilon(\omega(\tau))$.

6. Results and conclusions

All of the presented calculations and graphs show that the optimal angle of the nose inclination is $\phi = \arctan(5^{-1/2}) = 0.4205343352$ rad = 24° 5′ 41.43″. If the angle ϕ is less than the optimum value, then the amplitude of the resulting oscillations will be low and gears will be required. This will lead to a significant increase in the price of the whole pin. Otherwise, when the value of the angle ϕ is higher than the optimum value, there is a sharp increase in the instantaneous angular acceleration of the oscillations. High angular acceleration values mean that the entire pin must transmit high torque values to which the entire pin must be dimensioned. This will mean either increasing the size of the pin or using materials with satisfactory mechanical properties. This solution again leads to an increase in the price of the whole pin. If the pin is not appropriately sized for the transmitted torque, it may be mechanically damaged or blocked. The condition for the optimum angle value ϕ in Eq. (12), thus, appears to be the optimal compromise between the oscillation amplitude and the mechanical stress of the entire joint.

Of course, the optimum shape of the oscillations should have a sinusoidal waveform, i.e., $\theta(t) = a \sin(2\pi t/T)$, where θ = oscillation angle, t = time, a = oscillation amplitude, and T = oscillation period. The corresponding dependencies of angular velocity and angular acceleration over time are also in the form of sine or cosine

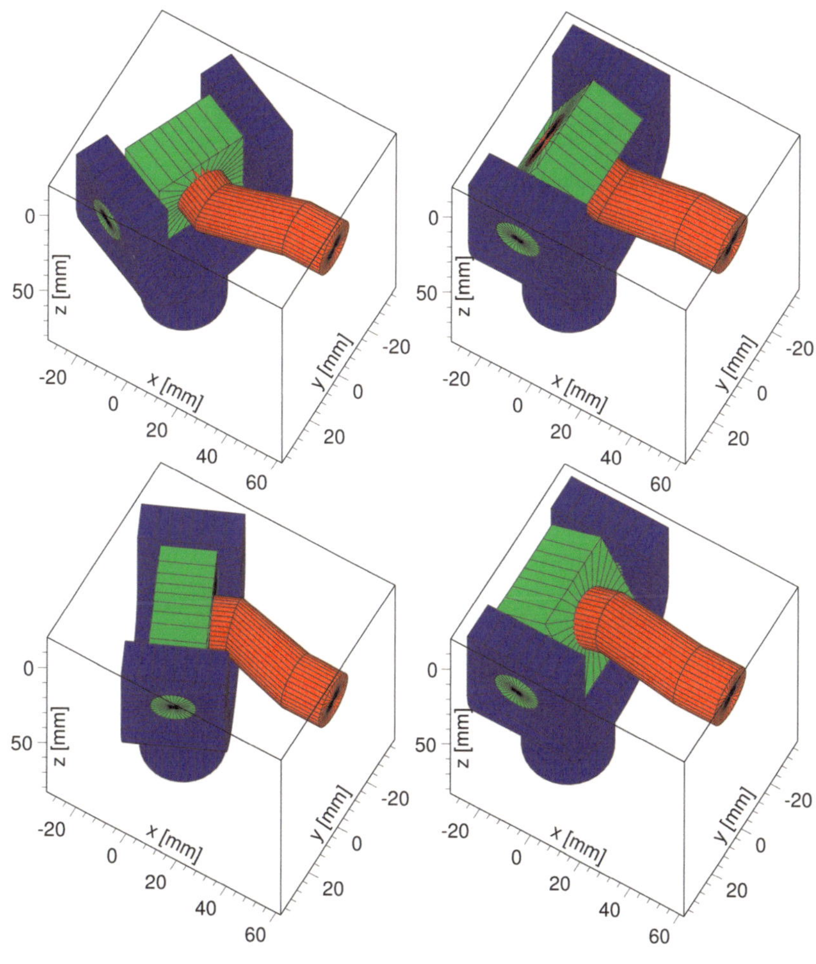

Figure 13.
Positions of components for rotation angles ψ = 0, 90, 180, and 270°.

functions. If we draw phase diagrams for these oscillations, then the graph of the dependencies $[\theta(\tau), \omega(\tau)] = \omega(\theta(\tau))$ and $[\omega(\tau), \varepsilon(\tau)] = \varepsilon(\omega(\tau))$ is the shape of an ellipse and the dependency $[\theta(\tau), \varepsilon(\tau)] = \varepsilon(\theta(\tau))$ is the shape of a line segment. The amplitude of the oscillations is then equal to the length of the main half-axis of the ellipse, which displays $\varepsilon(\theta(\tau))$.

If we examine the graphs in **Figures 10–12**, we find that the highlighted curve that corresponds to the optimum angle ϕ are an excellent compromise between the shape of the curve (ellipse or line segment) and the dimensions of the curve that correspond to the oscillation amplitude. Similarly, the graphs of functions $\theta(\tau)$, $\omega(\tau)$, and $\varepsilon(\tau)$, which correspond to the optimal value of the angle ϕ, are the best compromise between the desired function shape and the oscillation amplitude, see the graphs in **Figures 7–9**.

Figure 1 shows the entire inclined swinging pin, while **Figure 2** shows the tenon plotted for this angle. Several bottom views of the entire joint during one quarter of the working period are shown in **Figure 13**.

The main advantages of an inclined swinging pin compared to similar mechanisms are:

1. **Simplicity**: The whole pin consists of only three main components that can be forged.

2. **Compactness**: All components are similar in dimensions.

3. **Durability**: All parts of the mechanism are subjected to approximately the same forces. No component will receive more wear than others.

For these reasons, it is used in a wide variety of agricultural machines.

An animation (An_Joint.avi) that displays the entire device's activity can be downloaded from the Videos section.

7. Videos

All videos referenced in this chapter are available from: https://bit.ly/2vZVGXR

Author details

Stanislav Barton
Faculty of Electrical Engineering, Automatics and Informatics, Department of Informatics, Technical University of Opole, Poland

*Address all correspondence to: s.barton@po.opole.pl

References

[1] Horton HL, Newell JA. Ingenious Mechanisms. New York: Industrial Press Inc.; 1967;**4**:245-265. ISBN: 9780831110321

[2] Horton H. Ingenious Mechanisms. Volume IV [Internet]. 2017. Available from: https://books.industrialpress.com/ingenious-mechanisms-vol-iv-ebook-only.html [Accessed: March 15, 2019]

[3] Leon S. Linear transformations. In: Leon S, editor. Linear Algebra with Applications. 8th ed. New Jersey: Pearson Prentice Hall; 2010. pp. 166-198. ISBN: 139780136009290

Chapter 2

The General Kinematic Pair of a Cam Mechanism

Jiří Ondrášek

Abstract

At present, there are still increasing demands on the performance parameters of machinery equipment as well as cam mechanisms that belong to it. For this reason, the operating speeds and hence inertial effects of moving bodies, which limit the utilizable working frequency of machines, are increasing. These facts are the cause of higher wear and a decrease of the overall lifetime and reliability of machines. The force ratios in the general kinematic pair created by contact between the cam and the follower cause the contact stress. The generated stresses are transient and have a pulse shape. Fatigue damage of the cam working surface or the follower working surface may occur after exceeding a certain limit value of these stresses during the cam mechanisms running. This damage is in the form of cavities (pitting), which develop from cracks on the working surface. The chapter aim is to outline the issues of the dynamic stress of a general kinematic pair of a cam mechanism. One of the possible methods of the complex solution of the stress of the general kinematic pair is to use the possibilities of the finite element method in combination with the knowledge and conclusions of the contact mechanics.

Keywords: cam mechanism, cam, follower, general kinematic pair, contact stress

1. Introduction

Cam mechanisms are one of the basic objects in the design of production machines and equipment, whose characteristic feature is a high degree of automation and optimization of production and work processes. These mechanical systems are characterized by the transmission of large load possibility at high speed and positional accuracy of the working member of the relevant machinery. Their application is mainly connected with the so-called hard automation, which is characterized by unchangeable or difficult to change operations of the given technical equipment. Their widespread use is known in manufacturing and handling machines of the manufacturing industry, with their dynamic effects and properties greatly affecting the overall behavior, operation, and efficiency of such machinery. At present, increasing demands are placed on the performance parameters of such machinery. Therefore, the operating speeds and thus the inertial effects of the moving bodies are increased, thereby reducing the usable operating frequency of the machines. These facts cause greater wear and reduce overall machine lifetime and reliability and must be taken into account when designing and developing

 IntechOpen

them. The development of computer technology, numerical mathematics, and informatics enables the use of analytical and numerical methods in the design, development, and construction of cam mechanisms.

A general cam mechanism is typically referred to as a three-link mechanism with a single degree of freedom, which consists of two moving members mounted on a fixed frame (1). The moving members are the cam (2) and the follower (3) (see an example of a cam mechanism in **Figure 1**). In the case of the general cam mechanism, the general kinematic pair is formed of the contact of the working surfaces of the cam and the follower. The contact strain of the surface of the contact areas of the said type of kinematic pair, and in the vicinity of this surface, has a periodic course. As a result of the contact strain action, fatigue damage can occur on the contact surfaces. One of the criteria for such damage may be a value of the largest compressive principal stress in the contact area and in its vicinity. This is subsequently brought into the relationship with the strength limit of the respective material. The lifetime of the cam mechanisms is closely related to the choice of materials and their physical and mechanical properties from which the individual components are made, the method of material processing, and the technology of manufacturing of the individual parts or the way and the intensity of loading. It is connected with the way, intensity, and conditions of loading too. This issue connects the knowledge of theoretical, applied, and contact mechanics, tribology, material engineering, and structural analysis, though this theoretical knowledge must be supported by results from experimental identification.

In technical practice, multi-body cam systems are often encountered in addition to the basic three-body cam systems, which may also contain transforming linkages and mechanisms with constant gear ratio. Such mechanical systems are termed cam hinge or also combined cam mechanisms. These mechanical systems with a single degree of freedom are characterized by a uniform motion of the driving member, which generally does not need to be a cam, in most of their applications. The shape of the working surface of the cam realizes the working motion of the working member of the system, which is not necessarily a follower. In many cases, the kinetostatic analysis method is sufficient for the basic determination of the time course of the dynamic behavior of these mechanical systems. The kinetostatic solution determines the driving force effects, reactions in joints, and the force effects transmitted by the cam mechanism and linkages. For the solution, the knowledge is necessary of the kinematic quantities and geometrical mass parameters of all members of the mechanism as well as action force effects acting on

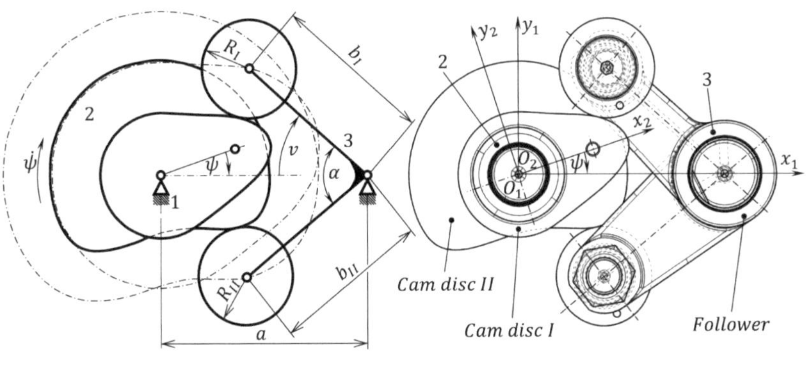

Figure 1.
A conjugate cam mechanism with an oscillating roller follower.

individual bodies. The results of kinetostatic analysis thus become the basic data for determining the distribution of contact stress in the contact area of the cam and follower.

2. Cam mechanisms

This section gives only basic information on how to solve the tasks of combined cam mechanisms. Fundamental terms and methods of analysis for the investigation of this type of the mechanical systems are specified. For example, the results of the kinematic analysis and synthesis are further used in kinetostatic or dynamic analysis tasks of these mechanical systems. The main goal of these analyses is to determine the courses of reactions in kinematic pairs, force loading of the cam and the follower, force effects acting on individual members, driving effects needed in the systems' movement, etc. Knowledge of the acting forces in the cam mechanism general kinematic pair is significant to determine the contact stress distribution in the contact of the cam and the follower.

Detailed knowledge of this issue may be found in [1, 2].

2.1 General cam mechanism

By definition, a general cam mechanism refers to a three-link mechanical system with a single degree of freedom that contains at least one cam linked with other members by means of at least one general kinematic pair. In this case, the general kinematic pair is formed by contacting the cam and the follower, whose movement is translation, rotation, or general. The cam mechanisms can implement a required working motion within a very precisely prescribed path with the use of a small number of bodies housed inside a relatively small space. The cam is the driving (or also input) member; in terms of shape, it is possible to define the basic cam types: radial, axial (cylindrical), and globoid. The follower is termed as the driven (or output or also working) member of a cam mechanism, which carries out the desired motion. The translating follower motion is defined as a translational or a general. The rotating follower, which performs a rotational motion, is usually called the lever. In order to reduce the passive resistance, the follower is often equipped with a roller in technical practice, whereby pure rolling in the interaction between the cam profile and the follower occurs, as is shown in **Figure 1**. This arrangement does not influence on the required follower motion.

One of the main conditions for proper operation of the cam mechanism is to maintain permanent contact of the follower with the cam during the action. This constraint of a general kinematic pair is achieved by a load or a redundant kinematic constraint. In the first case, the given contact is held using preloaded returnable compression springs, gravity forces, or inbuilt hydraulic or pneumatic elements. The disadvantage of this arrangement is the increased force loading and wear of the cam mechanism, which is caused by the preload required for the permanent contact between the follower and the cam. In the second case, contact by the redundant constraint is ensured by adding an extra linkage. For example, a grooved cam can realize such an arrangement. This embodiment is simple, but its disadvantage is the change in the rotation direction of the roller in the cam groove during the relative movement between the roller follower and the cam. This phenomenon is caused by a change in the sense of the transmitted normal reaction between the roller and the cam groove, because the pole of relative motion changes during the cam mechanism operation. As a result, the working surfaces of the groove are more worn in the points of the change in the roller rotation. Dual cam and roller follower systems are a more preferred design of the cam mechanism with the redundant kinematic constraint, although this solution is more expensive and complicated to manufacture. The conjugate, complementary, or double-disc cam is one including dual radial discs, each in contact with at least two driven followers coupled by a rigid or a kinematic linkage. The mobility of this mechanism is ensured by a special dimension arrangement, where the actions of both working surfaces of the dual cam must correspond exactly to each other. A schematic representation of a cam mechanism with radial conjugate cams and an oscillating dual roller follower is shown in **Figure 1**. The given constraint consists of another radial cam *II* and a roller follower *II*.

The shape of a cam contour is determined by the synthesis which is on the basis of the knowledge of the displacement law of the given cam mechanism and its dimensional parameters. The position of the cam relative to the frame of the cam mechanism is determined with an angular variable ψ, and the position of the follower is indicated with the generalized variable v (see **Figure 1**).

2.2 Combined cam mechanism

As a combined cam mechanism, it is generally called a mechanical system usually with a single degree of freedom which includes at least one general cam mechanism. This system usually also includes sets of various transforming linkages with not only a constant but also a generally variable gear ratio. They are most often complemented by simple linkages with lower kinematic pairs. In practice, lower pairs are generally planar couplings between two movable adjusted neighboring members. The members connected by a kinematic pair with the frame are referred to as the basic members of the transforming linkages. They perform rotational or translational motion. The basic representatives of such mechanisms are four-bar mechanism, crank mechanism, oscillating mechanisms, gears, etc. (see examples in **Figure 2**). The following findings are presented in accordance with the knowledge in publication [1, 3, 4]:

The input of the relevant linkage is an ordered triple of variables $\boldsymbol{\sigma}_k = (\sigma_k, \dot{\sigma}_k, \ddot{\sigma}_k)$ expressing motion of the input link of a mechanical system. The output is a triple of variables $\boldsymbol{\vartheta}_k = (\vartheta_k, \dot{\vartheta}_k, \ddot{\vartheta}_k)$, which represents the motion of the output link of the same system. Index k denotes the numerical indication of the relevant linkage, and it will be neglected in the next part of the text. The general equation of the linkage may be defined as an implicit function, in which the positional magnitudes σ, ϑ of basic members are time-related:

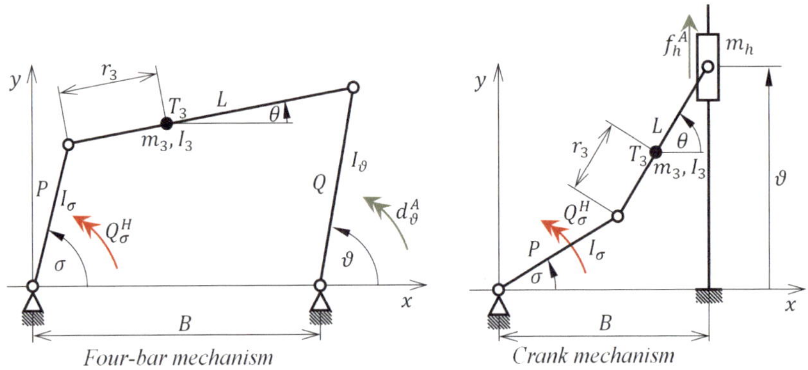

Figure 2.
Some common types of transforming linkages.

$$F(\sigma, \vartheta) = 0, \quad \sigma = \sigma(t), \quad \vartheta = \vartheta(t) \tag{1}$$

By differentiating Eq. (1) in time, a relation may be obtained between the velocity and the acceleration:

$$\frac{\partial F}{\partial \sigma}\dot{\sigma} + \frac{\partial F}{\partial \vartheta}\dot{\vartheta} = 0, \quad \frac{\partial F}{\partial \sigma}\ddot{\sigma} + \frac{\partial^2 F}{\partial \sigma^2}\dot{\sigma}^2 + 2\frac{\partial^2 F}{\partial \sigma \partial \vartheta}\dot{\sigma}\dot{\vartheta} + \frac{\partial F}{\partial \vartheta}\ddot{\vartheta} + \frac{\partial^2 F}{\partial \vartheta^2}\dot{\vartheta}^2 = 0 \tag{2}$$

Differentiation with respect to time is denoted by dots in Eq. (2). The linkage ratio is called a magnitude, which is dependent on the position of the linkage, and it is given by Eq. (3):

$$\frac{d\vartheta}{d\sigma} = -\frac{\partial F}{\partial \sigma}\left(\frac{\partial F}{\partial \vartheta}\right)^{-1}, \quad \frac{d\sigma}{d\vartheta} = -\frac{\partial F}{\partial \vartheta}\left(\frac{\partial F}{\partial \sigma}\right)^{-1} \tag{3}$$

The derivatives of Eq. (3) with respect to positions are given by Eq. (4):

$$\begin{aligned} \frac{d^2\vartheta}{d\sigma^2} &= -\left[\frac{\partial^2 F}{\partial \sigma^2} + 2\frac{\partial^2 F}{\partial \sigma \partial \vartheta}\cdot\frac{d\vartheta}{d\sigma} + \frac{\partial^2 F}{\partial \vartheta^2}\left(\frac{\partial \vartheta}{\partial \sigma}\right)^2\right]\left(\frac{\partial F}{\partial \vartheta}\right)^{-1}, \\ \frac{d^2\sigma}{d\vartheta^2} &= -\left[\frac{\partial^2 F}{\partial \vartheta^2} + 2\frac{\partial^2 F}{\partial \sigma \partial \vartheta}\cdot\frac{d\sigma}{d\vartheta} + \frac{\partial^2 F}{\partial \sigma^2}\left(\frac{\partial \sigma}{\partial \vartheta}\right)^2\right]\left(\frac{\partial F}{\partial \sigma}\right)^{-1} \end{aligned} \tag{4}$$

Introducing Eqs. (3) and (4) into Eq. (2) gives:

$$\dot{\vartheta} = \frac{d\vartheta}{d\sigma}\dot{\sigma}, \quad \ddot{\vartheta} = \frac{d\vartheta}{d\sigma}\ddot{\sigma} + \frac{d^2\vartheta}{d\sigma^2}\dot{\sigma}^2, \quad \dot{\sigma} = \frac{d\sigma}{d\vartheta}\dot{\vartheta}, \quad \ddot{\sigma} = \frac{d\sigma}{d\vartheta}\ddot{\vartheta} + \frac{d^2\sigma}{d\vartheta^2}\dot{\vartheta}^2 \tag{5}$$

which expresses relations for calculating the velocity and the acceleration of the output member.

The individual linkage can be placed into so-called *chains* when the outputs of the given mechanism are also the inputs of the next mechanism. The algorithms of linkage solving are marked with transformation blocks T_k as shown in **Figure 3**.

In relation to solving the problems with the combined cam mechanisms, we have introduced the following nomenclature (designation) of positional and kinematic magnitudes of the main members. Variable τ is assigned to the driving (input) member of a combined cam mechanism. The position of the driven (output) member is indicated with variable w. As a driven member, we usually consider such a working member or a body whose values are the result of the solution. The cam position with respect to the frame of the general cam mechanism is determined with an angular variable ψ, and the position of the jack is denoted by the generalized variable v (see **Figure 1**). The procedure of calculating a combined cam mechanism is shown through a block diagram, which is formed from transform blocks T_k and transform block C (see **Figure 4**). Transform block C expresses the algorithms of the solution of the general cam mechanism. From the view of the structure of a block diagram, it is advisable to divide the transform blocks into three groups. They are identified by this term *chain*:

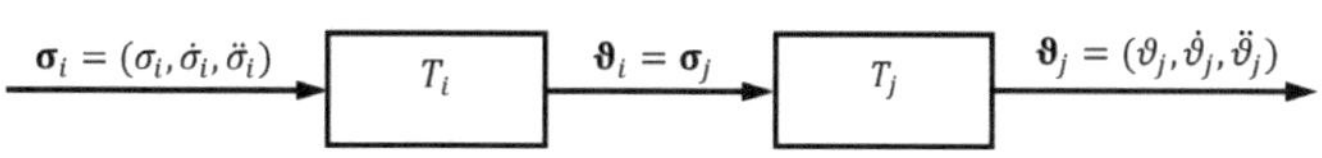

Figure 3.
A chain of transformation blocks of linkages.

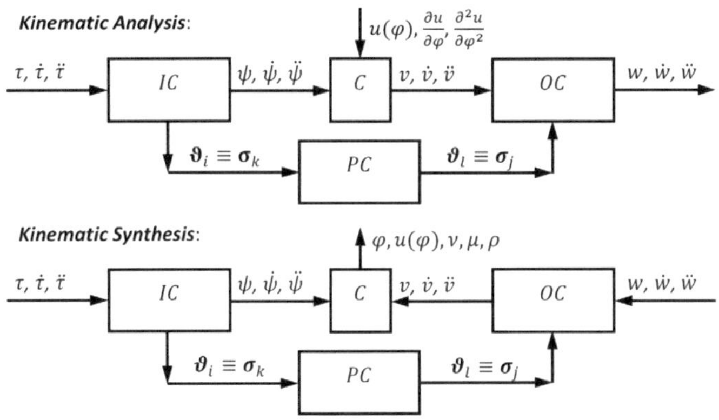

Figure 4.
Block diagrams of the computation of combined cam mechanisms.

- The input *chain IC* connects the input of the mechanical system to the input of block *C*.
- The output *chain OC* connects the output of block *C* to the output of the mechanical system.
- The parallel *chain PC* consists of blocks that do not belong to previous chains.

Indexes i, j, k, and l denote the $i^{\text{th}}, j^{\text{th}}, k^{\text{th}}$, and l^{th} linkage of the relevant chain.

The calculation of the positional and kinematic quantities of any member of a combined cam mechanism is designed as a kinematic analysis (see the block diagram in **Figure 4**). The input data are typical data (geometrical mass quantity and dimensions) on a mechanism, the procedure of computation of its chains, displacement function $\tau = \tau(t)$ of the driving member, and the shape of the cam theoretical profile $u(\varphi)$. The output of the solution is the displacement of the selected member of the mechanical system, which is expressed by a triple of variables $\mathbf{w} = (w, \dot{w}, \ddot{w})$. Kinematic synthesis of a combined cam mechanism is used to the computation of polar coordinates φ, u of a radial cam or cylindrical coordinates φ, x, y of an axial cam. Furthermore, via synthesis, normal angle ν, pressure angle μ, and radius of curvature of the cam profile ρ are set. The input data of the task include data on a mechanical system, the calculation procedure of its chains, displacement function of driving member $\tau = \tau(t)$, and driven member of the system $w = w(t)$.

2.3 Displacement law

The shape of a cam contour is determined by the synthesis which is on the basis of the knowledge of a displacement law of the given combined cam mechanism and its dimensional parameters. The following observations on the displacement laws are presented in accordance with the knowledge in publication [1].

A function assigning time t to a position variable of a given member of a mechanical system is termed to a motion function $w = f(t)$ of that member. The motion of the driving member is thus described by the independent motion function $\tau(t)$ and the motion of the driven member by the dependent motion function $w(t)$. The displacement law expresses a functional dependency of the driven member motion on the driving member motion of a combined cam mechanism $w = f(\tau)$. Displacement laws $w(\tau)$ of mechanical systems with the rotating input member are periodical functions with a period of 2π. The period 2π may be divided into motion and dwell intervals. Displacements on each motion interval may be different to a maximum total rise W and an expression of the normalized form $\eta = \eta(\xi)$ where the displacement and the range are in unity (see **Figure 5**). The variables ξ_k and η_k of kth motion interval are in linear correlation with the original variables τ and w and can be expressed as

$$\xi_k = \frac{\tau - \tau_{0k}}{T_k}, \quad \eta_k = \frac{w(t) - w_{0k}}{W_k} \tag{6}$$

The initial point O_k of displacement on each motion interval is defined by the coordinates τ_{0k}, w_{0k}. The interval length of the independent variable τ is given by the magnitude $T_k > 0$, and the maximum lift is expressed by the magnitude $W_k \gtrless 0$. The relationship between the original and normalized derivatives is

$$\frac{dw}{d\tau} = \frac{W}{T} \cdot \frac{d\eta}{d\xi} = \frac{W}{T} \cdot \eta'(\xi), \quad \frac{d^2w}{d\tau^2} = \frac{W}{T} \cdot \eta''(\xi) \tag{7}$$

Derivatives of the unity displacement to ξ will be denoted by primes, and index k, indexing the motion period, is neglected. When the variable w is a function t, then the derivatives will have the form

$$\frac{dw}{dt} = \frac{W}{T} \cdot \eta'(\xi) \cdot \frac{d\tau}{dt} = \frac{W}{T} \cdot \eta'(\xi) \cdot \dot{\tau},$$

$$\frac{d^2w}{dt^2} = \frac{W}{T} \cdot \left[\frac{1}{T} \cdot \eta''(\xi) \cdot \left(\frac{d\tau}{dt}\right)^2 + \eta'(\xi) \cdot \frac{d^2\tau}{dt^2}\right] = \frac{W}{T} \cdot \left[\frac{1}{T} \cdot \eta''(\xi) \cdot \dot{\tau}^2 + \eta'(\xi) \cdot \ddot{\tau}\right] \tag{8}$$

For the solution of problems related to the kinematic analysis and synthesis of cam systems, it is possible to use a broad set of displacement laws in a normalized form. These include, for example, polynomial, trigonometric, and exponential displacements and cycloidal, parabolic, and goniometric displacements (see [1]).

The choice of displacement law greatly has an influence over the dynamic properties and behavior of high-speed mechanical systems, and it should meet the following basic criteria within the specified conditions:

- In relation to the desired motion of the mechanism, the acceleration inertia forces, momentum, and performance of the cam mechanism should be always as small as possible.

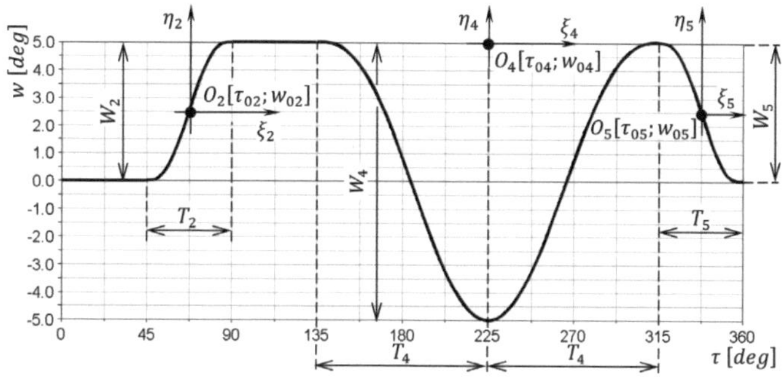

Figure 5.
An example of a displacement law.

- The vibrations forced by the movement of the mechanical system should be kept at a minimum.

Both criteria lead to low dynamic strain on the members of the mechanical system due to dynamic effects. In addition, the second criterion is related to the accuracy of adherence to the prescribed working member positions and the elimination of any noise sources. Comprehensive and detailed information on the issues of the displacement law choice is provided in [1].

3. Contact of cam and cam follower

In this text section, we will focus mainly on the general kinematic pair formed by contact of a cam and a roller follower. The mentioned type of kinematic constraint in the technical practice is usually most often constituted by a cylindrical roller and a cam or a crowned roller and a cam. In terms of computational purposes, we can substitute both mentioned contacts for the contact of cylindrical bodies with parallel axes and the contact of an elliptical body with a cylindrical one.

The contact area of the general kinematic pair is subjected to cyclic loading within the working cycle, while the contact surfaces are primarily in rolling contact in combination with a small percentage of mutual sliding. Thus, the transmission of normal and tangential forces is realized. These phenomena cause deformation of both bodies in the contact and cause contact stress in them. The state of stress on the working surfaces and under it is characterized by the principal stresses, which are transient and have the character of pulses with a period of 2π. Fatigue damage of the cam and follower contact surfaces may occur after a certain number of cycles in the operation of cam mechanisms, as long as a certain limit value of this stress is exceeded at any point of the contact area. This damage is in the form of cavities (pitting), which develop from cracks on the working surface. For cams with a hardened surface, this layer can be broken and then peels off (spalling) (see **Figure 6**). Both types of damage occur due to the contact stress that can be described by the theory of contact mechanics (see [5]). In terms of estimating the lifetime of the contact areas of the general kinematic pair, the distribution of the reduced stress in the surface areas and at a certain depth under it is therefore an important criterion. Thus, in the area of cam mechanisms, it is primarily a matter of determining the service life of the cam and follower contact surfaces depending on the conditions of their force loading.

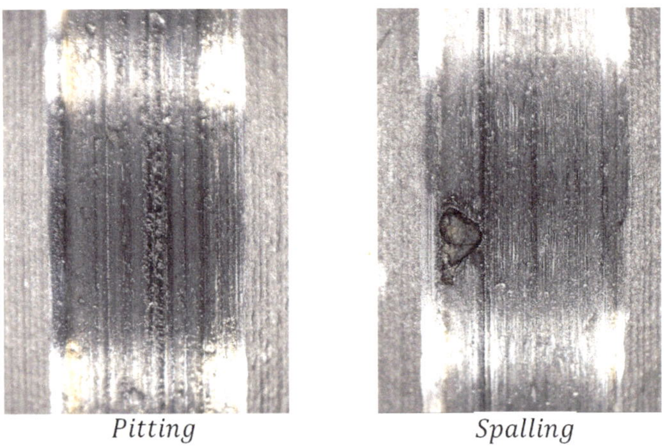

Figure 6.
Some common types of fatigue damages of cams and rollers.

The state of deformation and stress existing between the two elastic bodies in contact under load can be established both based on the contact mechanics and based on the use of the finite element method (FEM). The contact mechanics deals with the study of stress and deformation of solids being in contact at one point or along a line, acting under normal and also tangential forces. Physical and mathematical relationships are formulated on the basis of knowledge of continuum mechanics as well as mechanics of materials with the focus on elastic, viscoelastic, and plastic bodies in static or dynamic contact. The principles of contact mechanics are used to solve the problems of contact of rolling bodies (balls, rollers, barrels, needles, tapered rollers) and roller bearing rings, the contact of teeth in gearings, the contact of railway wheels and rails, mechanical constraints, and, last but not least, the contact of the cam and cam follower. The result of the calculations is also Hertzian contact stress, where there is local stress in the contact area, being caused by the contact of two curved areas, whereas these are slightly deformed due to the acting load. Hertzian contact stress is a fundamental quantity in formulating the equations for determining the carrying capacity and fatigue lifetime of cam mechanisms, bearings, gearings, and all objects in general, whose surfaces are in contact. Comprehensive and detailed information on the issues of contact mechanics is provided in [5].

3.1 Contact stress

When two three-dimensional bodies are brought into contact, they touch initially at a single point (*contact of the convex crowned roller with the cam*) or along a line (*contact of the cylindrical roller with the cam*). Under the action of the slightest load N, they will deform, and contact is made over a finite area which is small, compared with the dimensions of both bodies. During the compression distant points of the two bodies (T_1 and T_2) are approached the each other by a distance δ. Points S_1 and S_2 on the approaching contact surfaces are elastically displaced by amounts u_{z1} and u_{z2}, as shown in **Figure 7**. The shape of each surface in the contact region can be described by a homogeneous quadratic polynomial in two variables [5]:

$$z_i = \mp\frac{1}{2}\left(\frac{1}{\rho_{xi}}x^2 + \frac{1}{\rho_{yi}}y^2\right),\quad i = 1, 2 \tag{9}$$

where ρ_{xi} and ρ_{yi} are the principal radii of curvature of the surface at the rectangular coordinate system origin (see **Figure 8**). Contact stresses and deformations satisfy the differential equations for the stress and strain of homogeneous,

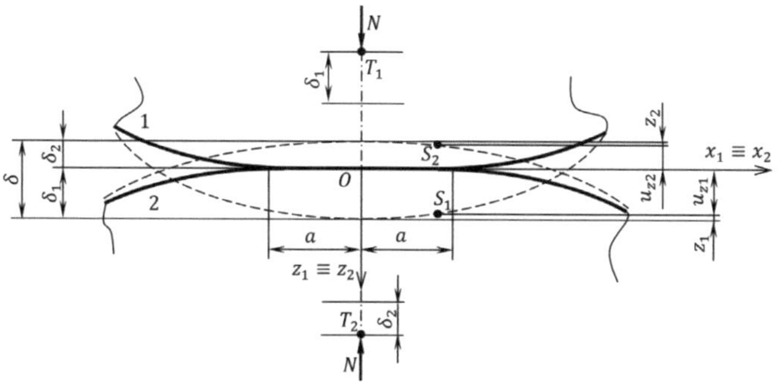

Figure 7.
The contact of two nonconforming bodies after elastic deformation.

isotropic, and elastic bodies in equilibrium. The pressure distribution on the contact area is given by the equation [5]:

$$p(x,y) = p_H\sqrt{1-\left(\frac{x}{a}\right)^2-\left(\frac{y}{b}\right)^2} \tag{10}$$

where a and b are respective major and minor semi-axes of the elliptical contact area and a maximum value p_H is called Hertzian pressure. The elasticity characteristics of the bodies in contact are introduced with the effective modulus of elasticity [5]:

$$\frac{1}{E^*} = \sum_i \frac{1-\nu_i^2}{E_i}, \quad i = 1, 2 \tag{11}$$

where E_i and ν_i are the respective Young's modulus of elasticity and Poisson's ratio of the individual solids.

The contact stress is highly concentrated in the vicinity of the contact area and decreases rapidly with an increasing distance from it. Thus, the stress area is close to the body contact. Since the contact surfaces are dimensionally small compared to the rest of the bodies, the stresses around the contact area are not too much dependent neither on the shape of the bodies in the contact nor on the way of mounting the bodies. This hypothesis simplifies the definition of boundary conditions and allows applying the theory of elasticity of large bodies.

In the case of the contact of *the convex crowned roller* with *the cam* and *the cylindrical roller with the cam*, it is clear that the abovementioned assumptions are satisfied. Thus, the results of Hertzian contact stress theory can be used to determine the stress state in their contact areas and at a certain depth under the surface (see [5]) or to use the results related to these contact types being given in publications [3, 6, 7], or we can directly calculate the contact stress for point or line contact by using the computational algorithms available on the webpage [8]. In **Figure 8**, there is a schematic presentation of the contact of the convex crowned roller and the cam (left) and distribution of the contact stress in the contact area (right). A similar case is shown in **Figure 9** with the only difference that there is the contact of the cylindrical roller and the cam. The stress state is in these cases expressed by principal stresses σ_x, σ_y, σ_z and reduced stress σ_{red}. These are universally compressive stresses, and their absolute value decreases with the distance from the surface. **Figure 10** generally describes the courses of individual stresses in the symmetry plane xz in dependence on depth z under the surface. From the course of the reduced stress σ_{red}, it is obvious that its maximum value is at a certain depth of z_e, which is expressed in unit form ξ_e. The principal stresses σ_x, σ_z reach their extreme

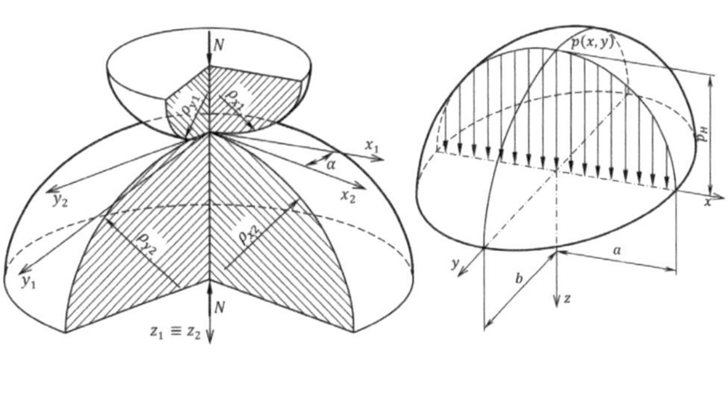

Figure 8.
Point contact.

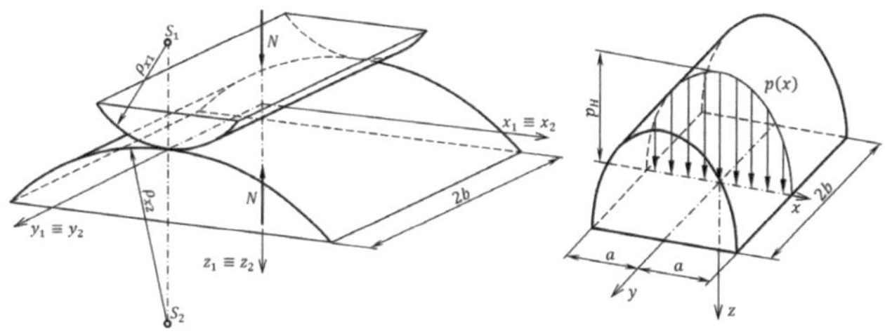

Figure 9.
Line contact.

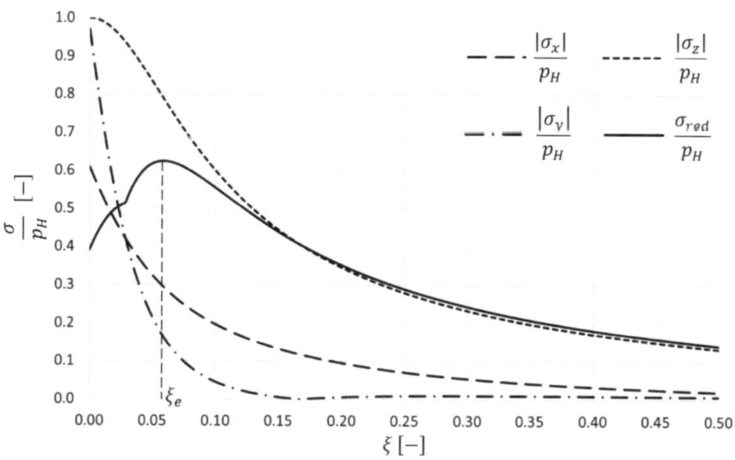

Figure 10.
The stress state in a symmetry plane xz.

values on the surface of the bodies in the contact area and equal to the Hertzian pressure value.

Another possible way of determining the stress in the contact areas of bodies is to use the finite element method. To achieve the relevant results, a dense finite element mesh in the contact area is required. This requirement leads to a large number of solved linear algebraic equations. The computational body contact algorithm is based on a numerical iteration of finding the elements in contact, so the numerical solution of the equations of the assignment takes place in several steps.

3.2 Reduced stress

In order to estimate the lifetime of the contact areas, it is first necessary to determine the magnitude of the reduced stress σ_{red}, which is compared with the value of the stress limit σ_h. The permissible stress determined on the basis of yield tensile strength $R_{p0.2}$ or ultimate tensile strength R_m refers to uniaxial stress. It is therefore necessary to have a criterion for comparing uniaxial stress and multiaxial stress in order to assess the strength at the multiaxial stress. These criteria provide hypothesis of strength. In the case of malleable materials, the failure occurs in the maximum shear stress plane at the tensile stress. The Guest's hypothesis is used (see, e.g., [9]). According to this hypothesis, the stress is judged from the point of view of the highest shear stress, which is proportional to the maximum of differences in the principal stresses:

$$\sigma_{red} = \max\{|\sigma_x - \sigma_y|, |\sigma_z - \sigma_x|, |\sigma_z - \sigma_y|\} \tag{12}$$

The principal stress components σ_x, σ_y, σ_z in the condition Eq. (12) are determined depending on the kind of contact of the general kinematic pair. These components must be established over the whole operation cycle of a cam mechanism to determine their maximum values. **Figure 10** shows that the highest value of the reduced stress is at a certain distance under the surface. In order to calculate the principal stresses, it is possible to use conclusions of contact mechanics for the respective contact of two elastic bodies, but it is also possible to utilize the method of finite elements (see [6, 7]).

In the case of general kinematic pairs, the contact load on the contact surface of the bodies and under it has the periodic course. At the contact areas of the general kinematic pair and below these points, contact stress becomes a periodical magnitude related to the angular cam displacement ψ. These transitory stresses are characterized by pulses with a periodicity 2π. The strength conditions acting under a variable loading are shown in Smith's diagram (see [10]), where the mean stress is determined by the equation:

$$\sigma_m = \frac{1}{2}[\max\sigma(\psi) + \min\sigma(\psi)], \quad \psi \in \langle 0, 2\pi) \tag{13}$$

and its amplitude by the equation:

$$\sigma_a = \frac{1}{2}[\max\sigma(\psi) - \min\sigma(\psi)], \quad \psi \in \langle 0, 2\pi) \tag{14}$$

In Eqs. (13) and (14), the variable ψ presents the rotation angle of the cam. Strength conditions under a variable load are given by the stress limit σ_h which is in agreement with the disturbance caused by the transitory stress. The reduced stresses σ_{red} (ψ, z) are limited by the actual strength condition, written as [1]:

$$\max\sigma_{red}(\psi, z) < \sigma_h, \quad \psi \in \langle 0, 2\pi), \quad z \geq 0 \tag{15}$$

where variable z is the depth under the contact surface of the cam or roller. Based on the tests of steels with ultimate tensile strength $R_m \in \langle 500, 1500\rangle MPa$, the fatigue limit of the material loaded by the tension is supposed as $\sigma_C \approx 0.35Rm$ and the yield strength as $R_{p0.2} \approx (0.55 \div 0.8)Rm$ (see [1]). Since the limit, the value of the transient stress, can be approximately determined according to $\sigma_h \approx \sigma_C \approx 0.7R_m$, then it is possible to replace the condition Eq. (15) by the inequality [1]:

$$\max\sigma_{red}(\psi, z) < R_{p0.2}, \quad \psi \in \langle 0, 2\pi), \quad z \geq 0 \tag{16}$$

Conditional inequality Eq. (16) describes the fact that during the operation of the cam mechanisms, no destructive action of elastic deformation occur in the general kinematic pair.

3.3 Lifetime of contact areas

The criterion of damage of the loaded contact areas of the cam and the follower in cam mechanisms is the formation of cracks and cavities, so-called pitting. The problem of the pitting formation on the surfaces of the bodies in contact is with that the initiation and propagation of cracks in the loaded material is completely unpredictable because the crack nuclei that form are distributed randomly in the material. Such crack nuclei are inclusions in material or surface irregularity caused by the production and treatment of this material. Therefore, it is difficult to predict exactly the stress state at the point of the contact area in which the damage occurs due to the load. This problem can be solved by introducing a criterion whereby the value of the highest principal compressive stress is determined in the contact area of the bodies, and this is brought into relation with the ultimate strength of the respective material. In the case of pure rolling contact, its magnitude is equal to the maximum value of the contact pressure—Hertzian pressure p_H.

In the previous part of the article, we briefly discussed ways of determining the stress in the surface and subsurface parts of the contact areas of the bodies. The following text will mention two theoretical approaches, one of which is presented in [1] and the other in [2]. Both procedures result from the knowledge of the contact mechanics, and based on this, the Hertzian pressure value p_H in the body contact area has been determined. Both methods require further the knowledge of some or several parameters that characterize the material from which the cam is made. These characteristic parameters are not available for all materials; therefore it is necessary to use experimental procedures. Not all materials are available for these materials, so further experimental procedures are required.

In the course of the operation cycle of the cam mechanism, no damage caused by the formation of pits is acceptable on the contact surface under load. Referring to [1], such fatigue damage will not occur if Hertzian pressure is given by the Niemann empirical relation in the form:

$$p_H \leq \frac{K}{N^{1/6}} f(H), \quad K = 4777 MPa \tag{17}$$

where the variable N is the lifetime in millions of cycles, H is the surface hardness in contact, and the function $f(H)$ represents the influence of the surface hardness on the permissible Hertzian pressure p_H (MPa). The factor K is a constant being determined empirically. The surface hardness H related to the maximum Hertzian pressure limit $p_{H\mathrm{Max}}$ in the course of the operation cycle is denoted as HB for the Brinell and HRC for the Rockwell scales. Then, for the function $f(H)$, the following empirical relations are formulated:

$$f(H) = \frac{HB}{1000}, \; f(H) = 0.251 + \frac{HRC}{100}\left[\frac{HRC}{100}\left(2.74 - 1.22\frac{HRC}{100}\right) - 0.6\right] \tag{18}$$

This criterion Eq. (18) is very simple because it is dependent on the only material parameter H.

Based on conclusions introduced in [2], a condition for the level of stress can be derived, in which the contact surface of the body will not be damaged:

$$p_H \leq \sqrt{\frac{E^*}{145.03789 \cdot \pi}\left(\frac{10^{\zeta}}{N}\right)^{\frac{1}{\lambda}}} \tag{19}$$

in which the constant E^* characterizes material elasticity of bodies in contact that is defined by Eq. (11). In Eq. (19) the strength factors ζ and λ express the slope and the intercept with the stress axis of the so-called *S-N* diagram in logarithmic coordinates for the surface fatigue limit of the respective material. *S-N* diagram is determined by regression analysis from a large number of test data.

Both conditions according to relations Eqs. (17) and (19) are illustrated graphically for selected steel C22 (1020, 1.0402) in **Figure 11**. The graphical

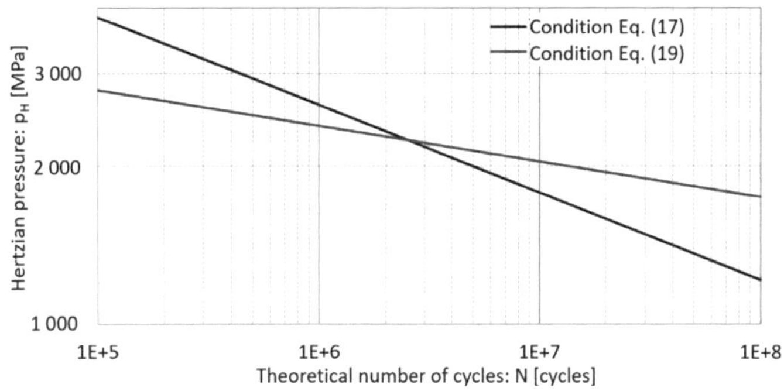

Figure 11.
Load-life relationships for steel C22.

representation shows a considerable difference between the mentioned theories in determining the lifetime of bodies in contact. In the region of limited lifetime, which is given by the number of cycles to damage $N = 10^7$ *cycles*, the conditions for the lifetime determination are according to condition Eq. (19) are usually more safety-related than condition Eq. (17). In the region of an unlimited lifetime, i.e., $N > 10^7$ *cycles*, it is the opposite case. Similar conclusions in both approaches to the determination of the lifetime of bodies in contact were also proved for other steels.

4. Effect of the roller crown shape on the cam stress

The geometrical shape of the roller crown itself has a significant effect on the stress distribution, due to the load and inertia effects in the contact areas of the general kinematic pair, which is usually formed in practice by a cylindrical roller and a cam or a crowned roller and a cam (see **Figure 12**).

It has been proven that the contact stresses in the vicinity of the shape, discontinuities in the contact area of the bodies in contact, are considerably higher than those reached outside the area of their immediate influence as described in the publication [2]. For this reason, the contact surfaces are more stressed, and their fatigue lifetime is reduced.

In general, we can expect to achieve a longer lifetime of the general kinematic pair of any cam mechanism by using the cylindrical profile of the roller crown. However, in the case of a conventional straight roller, there are discontinuities at the intersection of the roller cylindrical profile with the cam profile. These are caused by the fact that one contact area is axially shorter than the other and also by

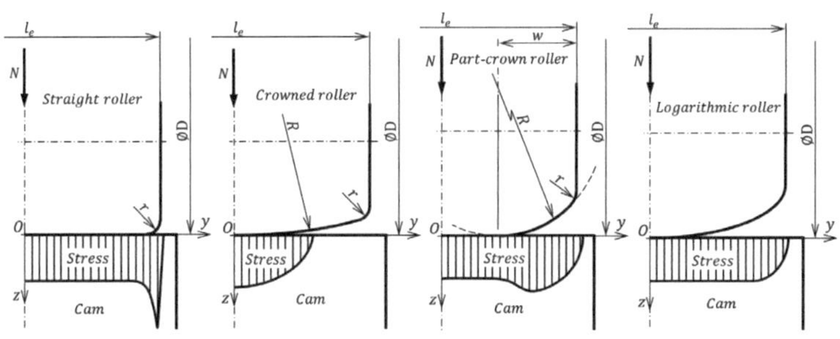

Figure 12.
Some types of roller profiles in contact with a cam.

chamfering the roller edges. In the vicinity of those profile discontinuities, the contact between the roller and cam cannot be considered as simply a straight line contact but for a more complex three-dimensional contact type. Therefore, in this case, it is not possible to apply the conclusions of Hertz's theory of contact to the calculation of the contact stress distribution. These discontinuities cause a high concentration of contact pressure at the appropriate point in the contact area. In fact, these local increases in the distribution of contact pressure can exceed the strength limit of the given material and thus cause plastic deformation in the contact area, the formation of residual stresses in the material, or hardening of steel. The area in question will further be more prone to fatigue damage to the contact surfaces such as pitting or spalling.

To ensure a more even distribution of contact stress in the contact area of the roller and cam, it is necessary to modify the shape of the axial cross-section of the roller crown. This is one of the reasons for the practical application of crowned rollers, when the radius of curvature of the crown profile is far greater than the radial radius of the roller (see **Figure 12**). Furthermore, we can easily compensate with their use misalignment between the roller and cam, without causing a fundamental change in the contact stress distribution. However, the largest concentration of contact stress distribution is achieved in the middle of the contact area with all the consequences as in the case of the straight roller.

The reduction of excessive contact stress in the vicinity of straight roller edges is achieved by such a shape of the axial cross-section of the crown, which includes straight line and tangent circular arcs that are connected to it on each side (see **Figure 12**). However, this shape of the roller crown leads to a certain concentration of contact stress in the transition from the cylindrical segment of the roller to the crowned one. Uniform distribution of stress for different levels of loading of the general kinematic pair contact areas is achieved by the logarithmic profile of the roller crown (see [11, 12]).This type of the roller crown is characterized by a monotonously decreasing profile from its center to the edge according to the logarithmic function (see **Figure 12**).

These facts will be demonstrated on examples of cam contacts with a cylindrical, crowned, and part-crown roller, which were defined using the finite element method (see [3, 13]). The formation of the model of a general kinematic pair using the finite element method is based on assumptions on the basis of which Hertz's theory of contact of two elastic bodies is derived (see [5]). The main assumption is that the contact area is continuous and much smaller than the characteristic dimensions of the bodies in contact. Therefore, the stresses in the vicinity of the contact area are not so dependent on the shape of the bodies in contact nor on how these bodies are fixed. Furthermore, it is assumed that the contact stress is very concentrated in the vicinity of the contact region and rapidly decreases with an increasing distance from it. The region of stress acting is therefore in the vicinity of the contact of the bodies. Through these basic assumptions, the definition of boundary conditions is simplified, and the application of the theory of elasticity of large bodies with sufficiently small deformations is allowed.

When creating a finite element model of a general kinematic pair of a cam mechanism, the contact of the roller with the cam will be replaced by the contact of two segments of solids of revolution. One of the solids of revolution represents the roller with the desired forming profile, and the other with a cylindrical profile replaces the cam. The radius of the cylinder is identical to the radius of curvature of the cam at the point of its contact with the roller. For the purposes of the computational analysis, we will use one eighth of each of them, assuming the parallelism of the axes of both replacement solids (see **Figure 13**). In the O_{xz} and O_{yz} planes in

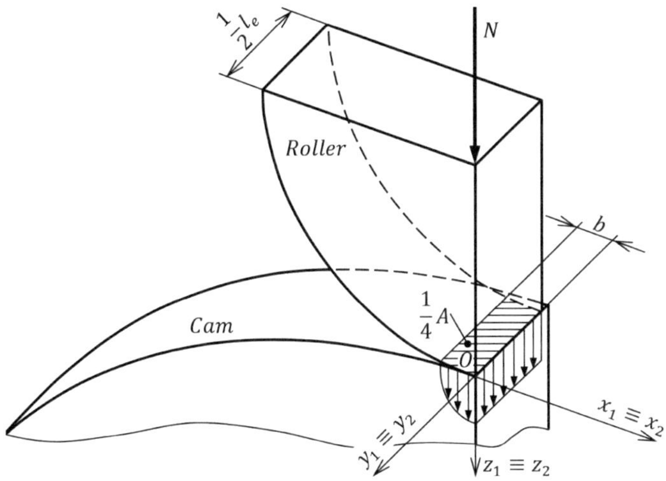

Figure 13.
A schematic drawing of a roller in contact with a cam.

		Straight roller	Crowned roller	Part-crown roller
Nominal diameter	D [mm]	35.0	35.0	35.0
Effective width	l_e [mm]	18.0	18.0	18.0
Fillet radius	r [mm]	0.6	0.6	0.6
Crown radius	R [mm]	—	500	200
Crown width	w [mm]	—	—	6.0

Table 1.
Nominal dimensions of rollers.

		Roller: 100Cr6	Cam: 16MnCr2
Young's modulus of elasticity	E [GPa]	210	206
Shear modulus	G [GPa]	81	79
Poisson's ratio	ν [−]	0.3	0.3038

Table 2.
Material characteristics of steels.

			Straight	Crowned	Part-crown
Load		N [N]	40,000	15,000	35,000
Hertzian pressure		p_H [MPa]	2500	2400	2400
Maximum reduced stress	Hertz theory	$\max(\sigma_{red})$ [MPa]	1440	1370	1480
	MKP		1970	1360	
Major contact radii		a [mm]	—	5.555	—
Half width of contact/minor contact radii		b [mm]	0.586	0.538	—
Depth of maximum reduced stress		z_e [mm]	0.42	0.38	0.38

Table 3.
Summary of results.

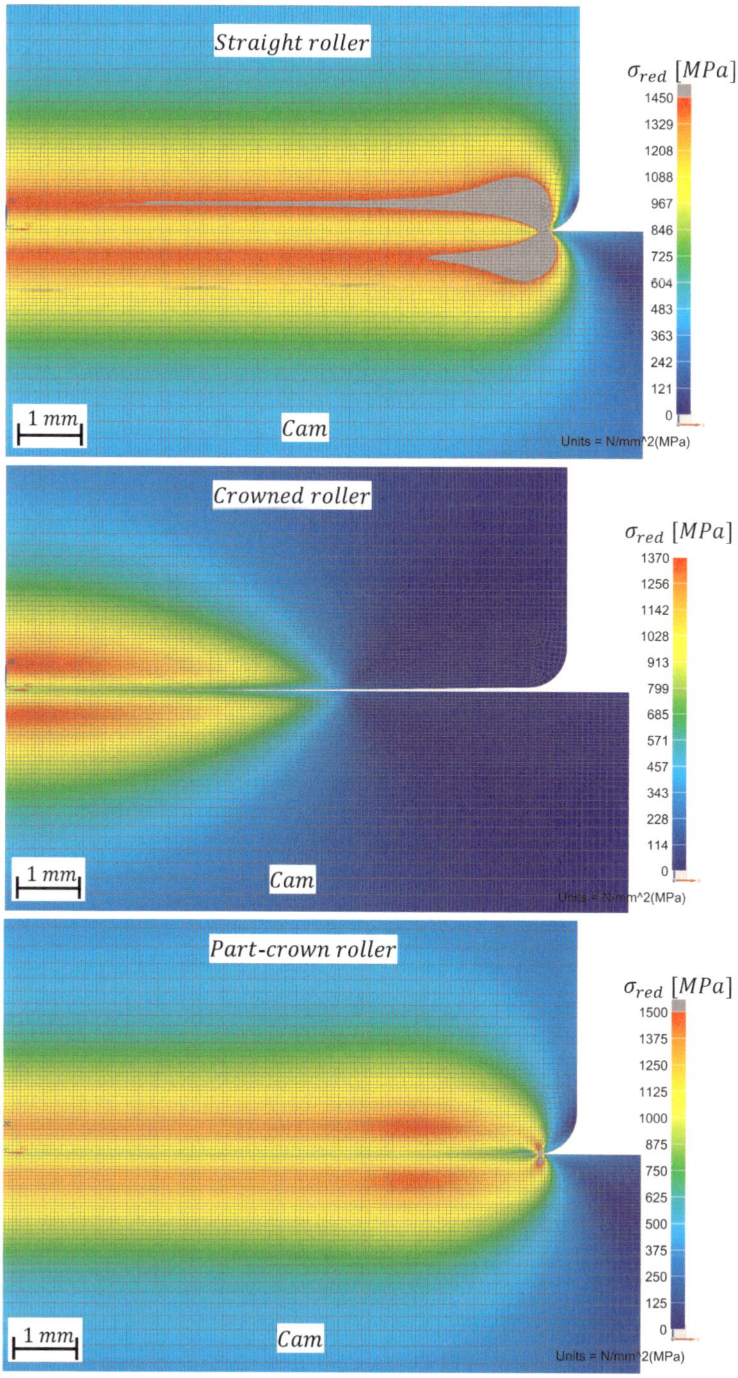

Figure 14.
The stress state in the symmetry plane xz determined by FEM.

accordance with **Figure 13**, there are then defined boundary conditions of the symmetry of the solved task accordingly. The lower surface of the body, which has the meaning of the cam, is fixed, i.e., all translations and rotations are prevented for all nodes of finite element mesh of the body in this area. Through the definition of the boundary conditions, the upper surface of the roller part is allowed to displace in the direction of the applied force N, i.e., in the direction of the z-axis of the *Oxyz* coordinate system. Also the acting of force *N* is evenly distributed over this area.

To achieve the relevant results, it is important that a uniform and dense finite element mesh be used to the discretization of the contact area and its vicinity of both bodies. The size of the elements can gradually increase with an increasing distance from the contact areas. For example, the size and shape of the contact area can be predicted by calculating based on Hertz's contact theory applied to contact of cylindrical bodies with parallel axes or to contact of a body with general profile and cylindrical body. Using this theory, we calculate the components of deformations and stresses in the contact area and its vicinity of both bodies in contact. Furthermore, the shape and size of the contact region are determined depending on the load size. This issue has been dealt with in articles [3, 6, 7] or is published in detail in [5]. Then, on the basis of the data thus determined, we define the space to create an acceptable finite element mesh density of the analyzed bodies with respect to the corresponding results compared to the real state.

The application of the above procedure will be demonstrated on the analysis of the contact stress distribution in the contact area of the cam, of which nominal dimensions are the width $l_{Cam} = 20$ mm and the curvature radius $R_{Cam} = 50$ mm in the vicinity of the contact point with the respective roller. We will consider three types of rollers, in which the nominal dimensions are listed in accordance with **Figures 12** and **13** and **Table 1**. The aim of the analysis of the contact stress of the cam is the determination of the maximum possible value of the reduced stress σ_{red} in the considered part of the cam depending on the greatest possible load of the roller by force *N*. Knowledge of this stress value is crucial with respect to fatigue damage of the working surfaces of the cam and the follower. The characteristic material parameters of the cam and the rollers are stated in **Table 2**.

Significant results are summarized in **Table 3**. Based on the size of the contact areas and the depth of the maximum reduced stress, the area was defined to create a quality network of elements in the vicinity of the contact of three roller and cam types in the creation of appropriate models using the FEM. In the case of cam and roller contact with the crown with convex segments, only the results from the FEM analysis are shown in the table. **Figure 14** shows the distribution of the reduced stress induced by contact of the said cam roller types with the cam. **Figure 15** shows the course of the maximum reduced stress in a depth of z_e according to the cam width for all rollers types in contact with the cam. Depth z_e expresses the distance from the contact area of the cam surface where the maximum value of the reduced stress is just obtained.

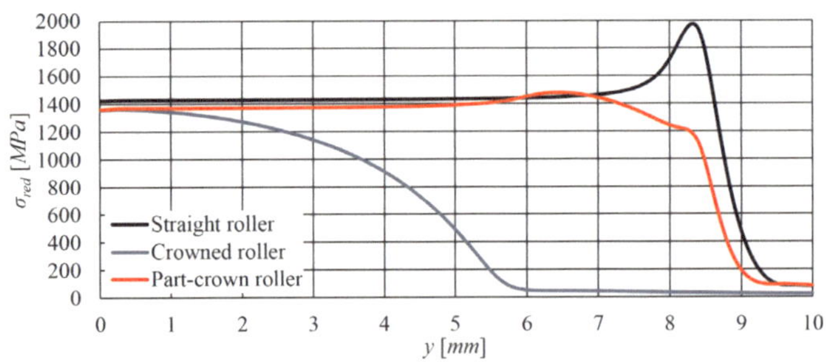

Figure 15.
The course of the maximum reduced stress depending on the cam width.

From **Table 3** and **Figures 14** and **15**, it is evident that there is an increase in the size of the reduced stresses in the vicinity of the profile discontinuities. This feature is particularly evident in the case of a cylindrical roller, where the contact between the roller and the cam cannot be regarded as merely straight but rather as a more complex three-dimensional type of contact. Therefore, Hertz's theory of contact cannot be applied to this type of contact around the shape discontinuities. Based on this theory, there are very good results compared to the FEM in contact of the general body with the cylindrical one and inside the contact area of the two cylindrical bodies with parallel axes. Furthermore, it is clear that a uniform distribution of stress can be achieved by such a shape of a roller crown whose profile includes the straight and two circular portions according to the schematic representation in **Figure 12**. This roller crown profile is advantageous in terms of load transfer capacity and process of its manufacture.

5. Conclusions

The presented chapter gives basic information on the stress problems related to a general kinematic pair of cam mechanisms. This type of kinematic pair is formed with at least one cam and a follower. The general cam mechanism is a very simple three-member mechanical system, which can implement the required working movements very accurately. Therefore, they are widely used in the design of various machines and equipment of the manufacturing industry. With the increasing pressure on the size and quality of machinery production of the manufacturing industry, the demand for its increased performance, reliability, and service life is growing. This fact is closely related to the detailed knowledge of the dynamic properties and behavior in the machinery during its operation. Thus, a dynamic response induced in the general kinematic pair is dependent both on the dynamic properties of all mechanical systems and on the prescribed displacement law. The choice of displacement law should be in conformity with the main requirements, which are, for example, reduced natural vibration, low dynamic load, high positional accuracy, and noiseless action.

Due to the effects of inertia and working forces, there are induced force ratios in the general kinematic pair that are the cause of contact strain. If a certain limit on this strain is exceeded, fatigue damage of the cam and follower contact surfaces may occur in the operation of the cam mechanisms. So knowledge of the distribution of the contact stress and its size are necessary when designing cam mechanisms. Contact stress expressed in Hertz pressure or the principal stresses becomes the criterion for determining the lifetime of the working areas of the mentioned kinematic pair. The lifetime itself depends on the choice of materials from which the individual parts are made and their physical and mechanical properties and the way of material processing and the production technologies of individual parts or the way and intensity of loading. There is currently a wide choice of materials for cams and cam followers of cam mechanisms. In technical practice, however, the most commonly used cams are made from steel. The surface of the cam or cam follower can be heat treated or chemically heat treated. The aim of the treatment is to achieve the desired mechanical or physical—chemical properties of the contact areas of the cam mechanisms. The purpose of this procedure is to increase the hardness and resistance of the contact surface against wear and to keep a resilient core of the respective component. The results from the abovementioned show that it is necessary to know various characteristic parameters describing material properties, heat treatment, or another technological processing for the lifetime estimation using theoretical methods. Therefore, experimental methods are an integral part of determining the lifetime of the working surface of the cam and the follower.

The working surfaces of the general kinematic pair of the cam mechanism frequently operate under extreme conditions, which are high loads, high sliding speeds, and poor lubricating conditions. Thus, this fact can lead to wear or excessive friction and thereby reduce the service lifetime and efficiency. These effects may be reduced by the application of coating on the working surfaces of the cam and follower. Coating is a technological process consisting of the fact that a very thin layer (the order of thousands of millimeters) is applied to the surface of an object, which has a relatively high hardness and strength compared to the underlying material. The thin layer of the coating forms a so-called barrier of the surface layers of the respective component against their chemical and physically mechanical wear. Coatings are generally used to improve hardness and tribological properties, wear resistance, and oxidation of exposed surface layers of the components. The coatings extend the lifetime of the sliding and rolling surfaces and help reduce the required power consumption while increasing performance. This decreases the use of lubricants and allows the use of new material combinations in the implementation of the relevant machinery. In some cases, coatings are even a necessary structural element for higher mechanical and thermal loads.

An effective way of analysis of the dynamic contact strain of the general kinematic pair of the cam mechanism is further presented here, which consists in the interaction of the knowledge and conclusions of Hertzian contact theory between two bodies and the advantages of using the finite element method. Using Hertz's contact theory, we predict the shape and size of the contact region of bodies in contact. In this way, we define the region to create an even and fine mesh of elements of an appropriate size in the vicinity of contact. Taking into account the assumptions of the Hertzian theory, the definition of the FEM model of a general kinematic pair is considerably simplified, and this reduces the number of algebraic equations needed to solve this problem; thereby, the computational time is reduced. This method leads to the achievement of relevant results compared to the real state. The method is presented in the task of determining the effect of the roller crown shape on the cam stress.

Acknowledgements

This chapter was created within the work on the TRIO II-FV20235 project—a project supported by the Ministry of Industry and Trade of the Czech Republic.

DOI: http://dx.doi.org/10.5772/intechopen.86682

Author details

Jiří Ondrášek
VÚTS, a.s., Liberec, Czech Republic

*Address all correspondence to: jiri.ondrasek@vuts.cz

References

[1] Koloc Z, Václavík M. Cam Mechanisms. 1st ed. Amsterdam: Elsevier; 1993. 424p. ISBN: 0-444-98664-2

[2] Norton R L. Cam Design and Manufacturing Handbook. 2nd ed. New York: Industrial Press, Inc.; 2009. 592p. ISBN: 978-0-8311-3367-2

[3] Ondrášek J. General Kinematic Pairs of Cam Mechanisms. 1st ed. Liberec: VÚTS, a.s.; 2018. 90p. ISBN: 978-80-87184-77-6 (in Czech)

[4] Ondrášek, J. The synthesis of a hook drive cam mechanism. In: Proceedings of the 6th International Conference on Modelling of Mechanical and Mechatronic Systems (MMaMS 2014); 25–27 November 2014; Vysoké Tatry/London: Elsevier Ltd; 2014. pp. 320-329. DOI: 10.1016/j.proeng.2014.12.129

[5] Johnson KL. Contact Mechanics. 1st ed. Cambridge: Cambridge University Press; 1985. 452p. ISBN: 0 521 34796 3

[6] Ondrášek J. The stress distribution in the contact region of a Cam mechanism general kinematic pair. In: Proceedings of the 4th MeTrApp Conference 2017 (MeTrApp 2017); 03–05 July 2017; Trabzon/Cham: Springer Nature Switzerland AG; 2018. pp. 99-108. DOI: 10.1007/978-3-319-60702-3_11

[7] Ondrášek J. The elastic compression in the contact region of a Cam mechanism general kinematic pair. In: Proceedings of the XII International Conference on the Theory of Machines and Mechanisms (TMM 2016); 06–08 September 2016; Liberec/Cham: Springer Nature Switzerland AG; 2017. pp. 57-63. DOI: 10.1007/978-3-319-44087-3_8

[8] Mesys AG. Contact Stress. 2010–2018. Available from: https://www.mesys.ag/?page_id=1220 [Accessed: April 23, 2019]

[9] Höschl C. Elasticity and Strength in Mechanical Engineering. 1st ed. Prague: SNTL/ALFA; 1971. p. 376. 04-021-71 (in Czech)

[10] Michalec J. Elasticity and Strength II. 1st ed. Prague: Publishing House CTU; 1994. p. 215. ISBN: 80-01-01087-2 (in Czech)

[11] Reusner H. The logarithmic roller profile the key to superior performance of cylindrical and taper roller bearings. Ball Bearing Journal. 1987;**230**:2-10

[12] Fujiwara H, Kawase T. Logarithmic profiles of rollers in roller bearings and optimization of the profiles. NTN Technical Review. 2007;**75**:140-148

[13] Ondrášek J. Effect of the roller crown shape on the cam stress. In: Proceedings of the 7th European Conference on Mechanism Science (EuCoMeS 2018); 04–06 September 2018; Aachen/Cham: Springer Nature Switzerland AG; 2019. pp. 223-230. DOI: 10.1007/978-3-319-98020-1_26

Section 2

Compliant Mechanisms

Chapter 3

Modeling and Design of Flexure Hinge-Based Compliant Mechanisms

Sebastian Linß, Stefan Henning and Lena Zentner

Abstract

A compliant mechanism gains its mobility fully or partially from the compliance of its elastically deformable parts rather than from conventional joints. Due to many advantages, in particular the smooth and repeatable motion, monolithic mechanisms with notch flexure hinges are state of the art in numerous precision engineering applications with required positioning accuracies in the low micrometer range. However, the deformation and especially motion behavior are complex and depend on the notch geometry. This complicates both the accurate modeling and purposeful design. Therefore, the chapter provides a survey of different methods for the general and simplified modeling of the elasto-kinematic properties of flexure hinges and compliant mechanisms for four hinge contours. Based on nonlinear analytical calculations and FEM simulations, several guidelines like design graphs, design equations, design tools, or a geometric scaling approach are presented. The obtained results are analytically and simulatively verified and show a good correlation. Using the example of a path-generating mechanism, it will be demonstrated that the suggested angle-based method for synthesizing a compliant mechanism with individually shaped hinges can be used to design high-precise and large-stroke compliant mechanisms. The approaches can be used for the accelerated synthesis of planar and spatial flexure hinge-based compliant mechanisms.

Keywords: compliant mechanism, flexure hinge, deformation behavior, motion behavior, modeling, design

1. Introduction

A mechanism is generally understood as a constrained system of bodies designed to convert forces or motions. Fulfilling the function of power transmission (drive in the actuator system) or motion transmission (guidance in the positioning system), mechanisms are typical parts of a mechatronic motion system. For the realization of high-precise motion, increasingly compliant mechanisms are used instead of rigid-body mechanisms. A mechanism that gains its mobility fully or partially from the compliance of its elastically deformable parts rather than from rigid-body joints only is named as compliant mechanism [1, 2].

In precision engineering and micromechanics, there are increasingly high requirements for the motion system—especially regarding the smoothness, resolution, and repeatability of the motion. Therefore, compliant mechanisms with concentrated or distributed compliance have become established for special positioning [3], adjustment [4], manipulation [5], or metrology [6] tasks. In these monolithic mechanisms, flexure hinges are mostly used as materially coherent revolute joints [7], while a high motion accuracy in the micrometer range can especially be achieved by common notch flexure hinges [8].

Nevertheless, the output stroke or motion range of such compliant mechanisms is considerably limited by the material strength since identical circular notch shapes are used for all hinges in the mechanism in most cases, even if they achieve different rotation angles. For high-strength metals, which are typically used for precision engineering applications, the rotation of flexure hinges is limited to small angles of a few degrees [9]. The demand for a larger angular deflection and a lower shift of the rotational axis results in numerous possible notch shapes and in a variety of sometimes very complex types of a separate flexure hinge, like the butterfly hinge [10]. Alternatively, mechanisms with a significantly increased hinge number in the kinematic chain are proposed to increase the range of motion, for example [11]. To further increase the stroke, often complex combinations of several substructure mechanisms are used in planar or spatial compliant stages, for example, reported in [9].

The sequential procedure including structural type synthesis, dimensional synthesis, and embodiment design, often used for rigid-body mechanisms, cannot be applied to compliant mechanisms straightforward, since the force/displacement limits of the flexure hinges must be matched with the required motion task. Thus, kinematic and kinetic behavior must be considered simultaneously for synthesis. Furthermore, the complex deformation and motion behavior of compliant mechanisms complicates both their accurate modeling and purposeful design. Hence, the synthesis is iterative, nonintuitive, and often time-consuming so far, and specific optimization approaches, for example [12], cannot be generalized. However, optimizing the shapes of easy-to-manufacture and mainly used notch flexure hinges may prove useful in the synthesis of compliant mechanisms. Among many possible notch shapes, power function flexure hinges, based on the higher order polynomial hinges suggested in [13], are especially suitable because they are highly variable and allow a simplified modeling, too [14].

In this chapter, a survey of different methods for the general and simplified modeling of the elasto-kinematic properties of flexure hinges and compliant mechanisms is provided for four certain hinge contours, the circular, the corner-filleted, the elliptical, and the power function-based contours, with different exponents. Based on nonlinear analytical calculations and FEM simulations, several approaches and guidelines like design graphs, design equations, design tools, or geometric scaling are presented which can be used for the flexure hinge design. The results are confirmed by means of analytical modeling and FEM simulation. The main approach with regard to the mechanism synthesis is to design each flexure hinge in a compliant mechanism individually in dependence of the known relative rotation angles in the rigid-body model. A four-bar path-generating mechanism is used as an example to show the benefits of the synthesis method regarding both a high precision and a large stroke in comparison to the use of identical notch geometries. Thus, the need for simulation is reduced.

2. Flexure hinge-based compliant mechanisms

A structural part of a compliant mechanism with a greatly increased compliance can be seen as a compliant joint, which allows at least one relative motion due to deformation, but it is normally limited to a localized area. In dependence of the form of the relative motion, three types for a joint with one degree of freedom ($f = 1$) are existing, the revolute pair, the prismatic pair, and the screw pair (see **Table 1**).

Conversely to rigid-body joints, in which two rigid links form either a form-closed or force-closed pairing, neighboring links of a compliant mechanism are connected to each other in a materially coherent way. Thus, an increased compliance can be achieved through a variation of geometry and/or a variation in material, while the geometric design is in the focus of the following investigations. In this chapter, macroscopic compliant mechanisms with flexure hinges realizing a desired rotary motion are regarded, since they are used in most cases.

2.1 Analysis and synthesis of compliant mechanisms

For the synthesis of a compliant mechanism, three main approaches are suggested in literature: synthesis through the rigid-body replacement method (e.g., [15]), synthesis through the topology optimization method (e.g., [16]), and synthesis through constrained-based methods (e.g., [17]). In order to realize a better guidance accuracy, the rigid-body replacement synthesis is more suitable than the topology optimization synthesis [18]. Therefore, here the purposeful design of a compliant mechanism based on the rigid-body model is meant by speaking of synthesis. The geometric design of the incorporated flexure hinges is a key point during the synthesis, because often multi-objective design criteria exist.

Regarding a four-bar Roberts mechanism realizing an approximated straight-line path of a coupler point P, the rigid-body mechanism and the compliant counterpart are shown in **Figure 1**. For the replacement, the same initial position of the compliant mechanism with the crank angle γ is used as in the rigid-body model. The Roberts mechanism with four hinges is a typical path-generating mechanism which is used for the rectilinear guidance of a coupler point in precision engineering applications, for example [19–22].

Joint type	Symbol	Rigid-body joint	Compliant joint
Revolute pair (hinge)	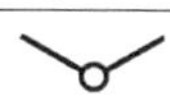	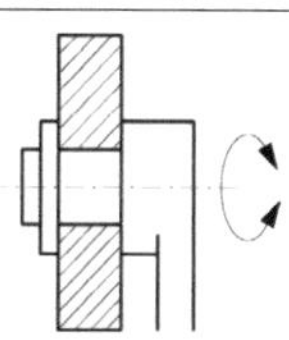	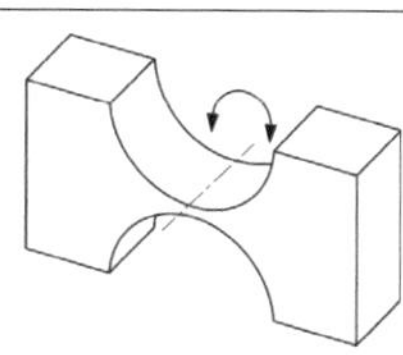
Prismatic pair	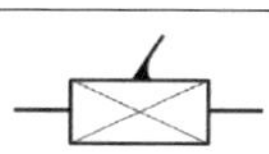	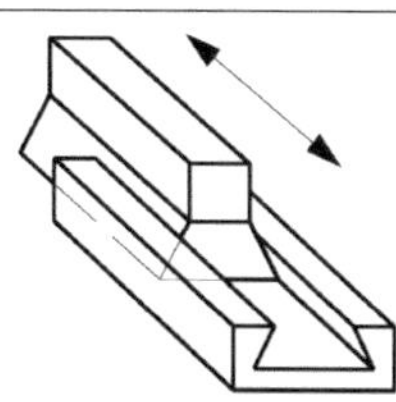	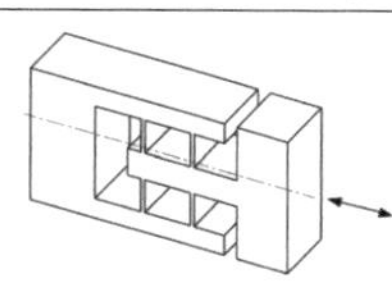
Screw pair	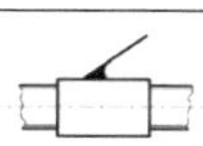	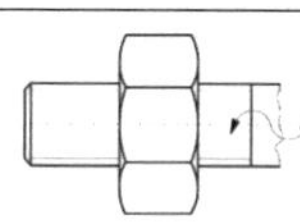	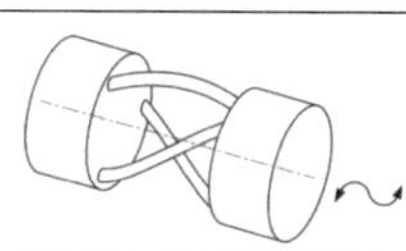

Table 1.
Classification of joints with f = 1 by means of the form of relative motion.

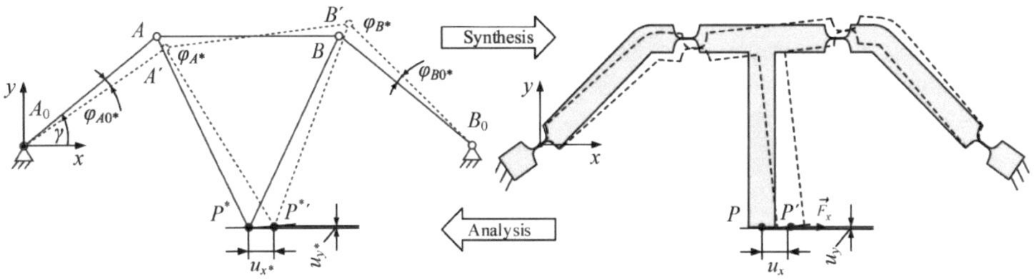

Figure 1.
Analysis and synthesis of a compliant mechanism based on the rigid-body model using the example of a path-generating Roberts mechanism.

In contrast to the synthesis, the analysis describes the modeling of the rotation axes and link lengths of the rigid-body model based on the compliant mechanism, for example [1, 2]. Additionally, the bending stiffness of all hinges has to be considered.

With a few exceptions (e.g., [5, 23]), almost identical flexure hinges are used in the same single compliant mechanism. However, the relative rigid-body-based rotation angles φ^* for the desired motion of the mechanism are different for the incorporated hinges in most cases. Due to the different rotation angles, different flexure hinge contours are also required. Because the deflection angle φ of each flexure hinge is approximately equal to φ^* [24], an angle-based and goal-oriented four-step synthesis method for using individually shaped notch flexure hinges in one compliant mechanism can be applied [25]. The basic phases are (cf. Section 4):

i. Synthesis of a suitable rigid-body model

ii. Replacement and design of the compliant mechanism

iii. Goal-oriented and angle-based geometric design of the flexure hinges

iv. Verification of results and proof of requirements.

In literature, the specific geometric design of the flexure hinges during synthesis is only considered when using almost identical hinges in a compliant mechanism and standard contours with a limited variability like corner-filleted hinges [26].

2.2 Classification of compliant mechanisms

In dependence of the existence of rigid-body joints, compliant mechanisms can be separated into the categories of fully compliant mechanisms or partially compliant mechanisms, while the presented design guidelines in this chapter are suitable for both. Additionally, fully and partially compliant mechanisms can be separated into mechanisms with concentrated or distributed compliance [2] (see **Table 2**), while mechanisms with concentrated compliance are regarded here.

Furthermore, the presented results in this chapter are focused on planar compliant mechanisms (see **Figure 2**). Nevertheless, the suggested methods and design approaches can be used for spherical and spatial compliant mechanisms, too.

2.3 Classification of flexure hinges

A flexure hinge is understood as a compliant joint which approximately acts as a hinge due to flexural bending. Thus, the form of relative motion can only be idealized as a rotation. Because of their monolithic arrangement, compliant joints

Mechanism	Fully compliant mechanism	Partially compliant mechanism
With concentrated compliance	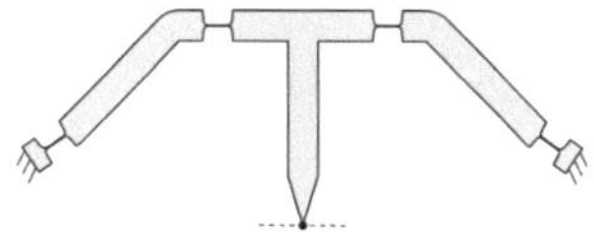	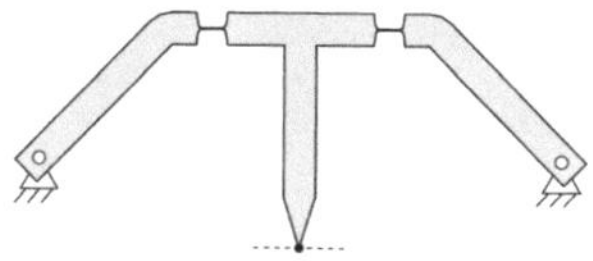
With distributed compliance	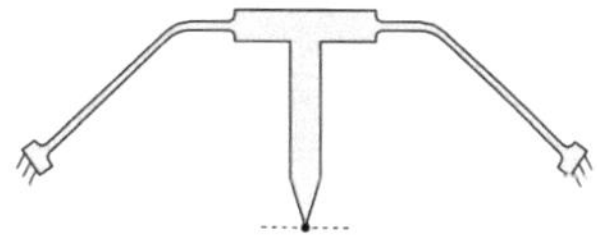	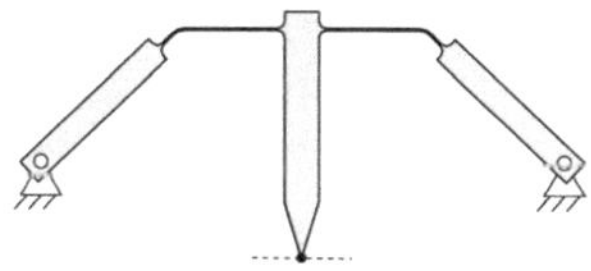

Table 2.
Classification of compliant mechanisms by means of the structural design and the distribution of compliance.

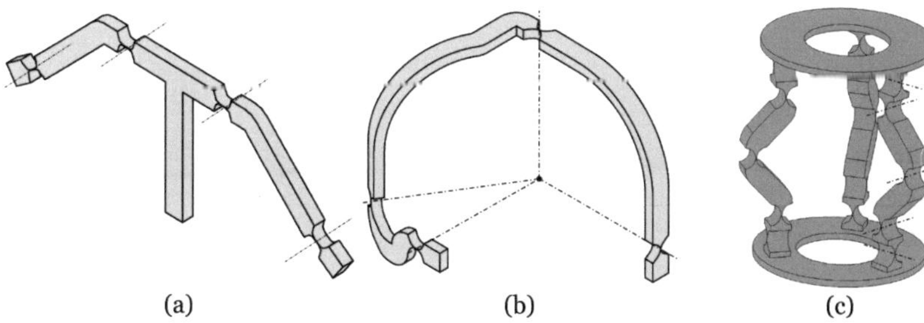

Figure 2.
Classification of compliant mechanisms by means of the position of the revolute axes: (a) planar mechanism, (b) spherical mechanism, and (c) spatial mechanism according to [27].

provide numerous approaches for the design of a flexure hinge. Based on the well-described leaf-type flexure hinge [28], many different flexure hinge types have been developed in the past decades or introduced in recent works in order to realize a larger angular deflection and/or a more precise rotation (see **Figure 3**) [10, 29–31]. Many more flexure hinge types are classified in [32].

The design guidelines in this chapter are focused on notch flexure hinges because different design goals can be met by selecting between comparable, simple notch hinge designs already, largely due to a great contour variety [32]. Due to their low complexity, they are easy to manufacture and therefore mainly used in compliant mechanisms, especially in kinematic chains with a higher link number.

Notch flexure hinges have often geometrically been designed so that various cutout geometries are proposed to describe the variable contour height. There are mostly predefined basic geometries which lead to the typical precise hinge with a semicircular contour, the large-deflective hinge with a corner-filleted contour, or the elliptical hinge as a compromise [33]. Furthermore, flexure hinges are designed with other elementary or complex geometries (e.g., [34]) to realize special properties. Higher order polynomial functions are not state of the art. But among the

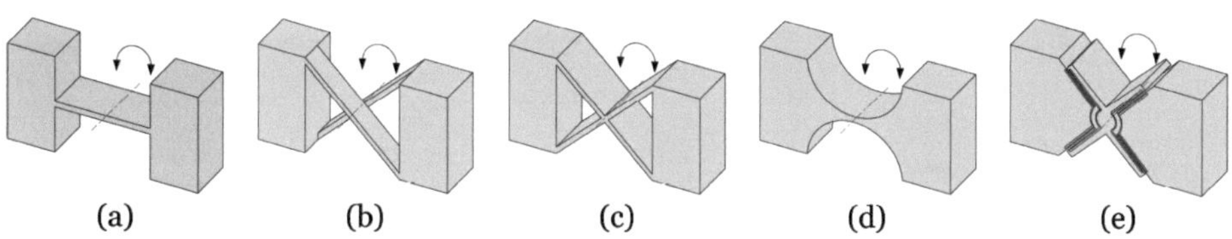

Figure 3.
Typical types of flexure hinges used to achieve a rotational motion with one degree of freedom (f = 1): (a) leaf-type hinge, (b) crossed leaf-type hinge, (c) prismatic crossed hinge, (d) notch hinge, and (e) multi-trapezoidal hinge/butterfly hinge.

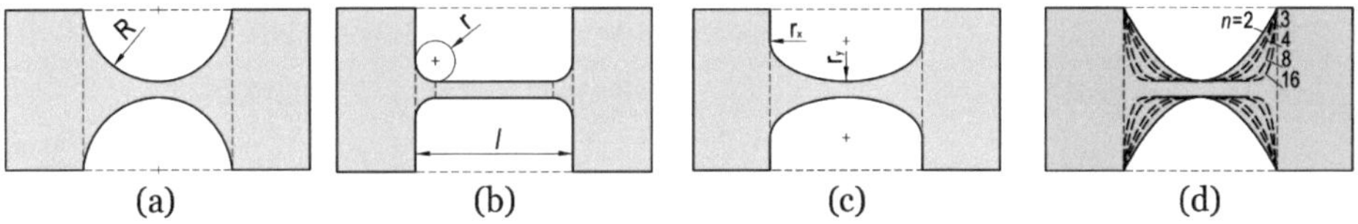

Figure 4.
Typical geometries of notch flexure hinges with their contour-specific parameters: (a) circular contour with radius R*, (b) corner-filleted contour with stress-optimal fillet radius* r = 0.1 l*, (c) elliptical contour with major axis* r_x *and minor axis* r_y*, and (d) variable power function-based contour, shown for different exponents* n*.*

variety of geometries, especially these contours offer high optimization potential, while a comparatively simple modeling is possible [35]. Thus, the advantages of the polynomial contour are implemented and extended to a power function contour to offer a wider range of possible hinges due to a rational exponent n [14], with

$$h_{\mathrm{c}}(x) = h + \frac{(H-h)}{\left(\frac{l}{2}\right)^{n}}|x|^{n}, \text{ with } n \in \mathbb{R}. \tag{1}$$

In the following, four certain flexure hinge contours are considered (see **Figure 4**): the semicircular contour, the corner-filleted contour with a stress-optimal and hinge length-related fillet radius $r = 0.1\,l$ [36], the semielliptical contour, and the power function-based contour. The remaining contour height functions and detailed information about the depicted segment-wise contour modeling are given in [14]. Many more notch geometries are classified in [32].

3. Modeling and design of notch flexure hinges

As a flexure hinge, a monolithic, small-length, and elastically deformable segment of a compliant mechanism with the variable and symmetric contour height $h_c(x)$ and a rectangular cross section is defined, which provides a relative rotation of two adjacent links mainly limited due to bending stress (see **Figure 5**). Since not only the notch segment undergoes deformation, it is recommended to model the hinge with little segments of both adjacent links and as a 3D solid structure [37, 38].

In the following, three important rotational performance properties are considered. A flexure hinge provides a restoring force (a property called *bending stiffness*), which may be beneficial in precision applications, too. Depending on the material

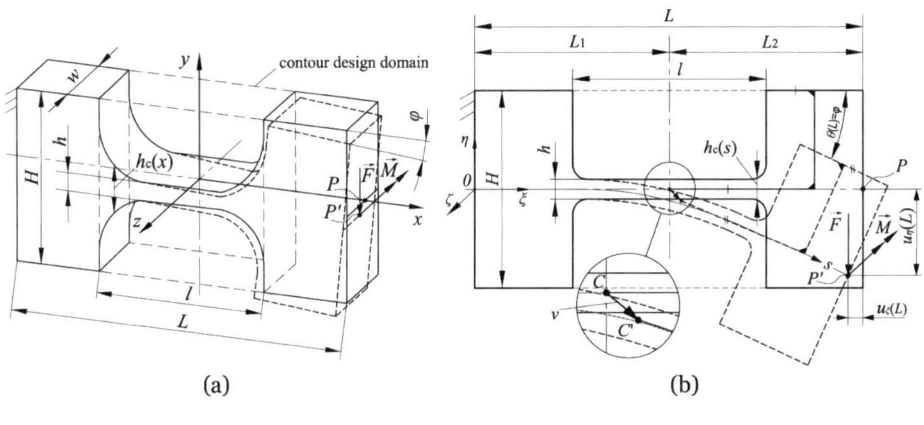

Figure 5.
Modeling of a notch flexure hinge under a bending moment and/or a transverse force load: (a) hinge with a variable hinge height within the contour design domain, the geometric parameters and the deflected state and (b) parameters for the theoretical characterization and approach for the definition of the rotational axis shift based on guiding the center with a constant distance, the fixed center approach.

coherence, the angular rotation of a flexure hinge is restricted by maximum acceptable stresses or elastic strains (*maximum angular deflection*). Therefore, the stroke of a compliant mechanism is limited as well through the joint with the largest deflection angle in the kinematic chain, assuming the same contours are used. In addition, no exact relative rotation is possible with a flexure hinge, because its axis of rotation is always shifted as geometric and load parameters vary (*rotational precision*) [36, 39]. In turn, this can lead to path deviations in the compliant mechanism compared with the rigid-body model that can no longer be considered negligible, especially in precision engineering applications [24, 40].

Regarding the influence on the flexure hinge properties, two main groups of geometric design parameters are existing, the hinge dimensions (L, l, H, h, w) and the hinge contour or notch shape (function $h_c(x)$), while the total height H represents the link height in the compliant mechanism, too. Hence, four geometric parameters—the hinge total length L, the hinge notch length l, the minimum hinge height h, and the hinge width w—can be varied within the design domain according to the following introduced dimensionless ratios:

$$\beta_L = \frac{L}{H}, \beta_l = \frac{l}{H}, \beta_h = \frac{h}{H}, \text{ and } \beta_w = \frac{w}{H}. \tag{2}$$

For a separate flexure hinge, it is known that the properties depend on the basic geometric dimensions as follows [41, 42]: the bending stiffness and maximum stress increase in particular as the minimum hinge height h increases and the rotational precision decreases with an increasing minimum height h. Furthermore, several different and partly contrary recommendations for some hinge dimension ratios of circular and corner-filleted flexure hinges are existing (cf. [32]).

Other than that, the high-strength aluminum alloy AW 7075 with Young's modulus E = 72 GPa, Poisson's ratio ν = 0.33, and the admissible elastic strain limit ε_{adm} = 0.5% is chosen as a typical material which has been used for multiple high-precision engineering applications, for example [3, 21].

3.1 Nonlinear FEM simulation

For the quasi-static structural FEM simulation, performed with ANSYS Workbench 18.2, the hinge is modeled as a 3D structure with Solid186 hexahedral elements. The CAD model and FEM model are shown in **Figure 6**. The FEM model is considered with a fixed support on one side, and it is free on the opposite side. The free end is stepwise loaded with a bending moment or a directionally constant transverse force applied at an edge parallel to z. The analysis of two points on the loaded hinge side enables an accurate calculation of the rotation angle φ. Hence, the

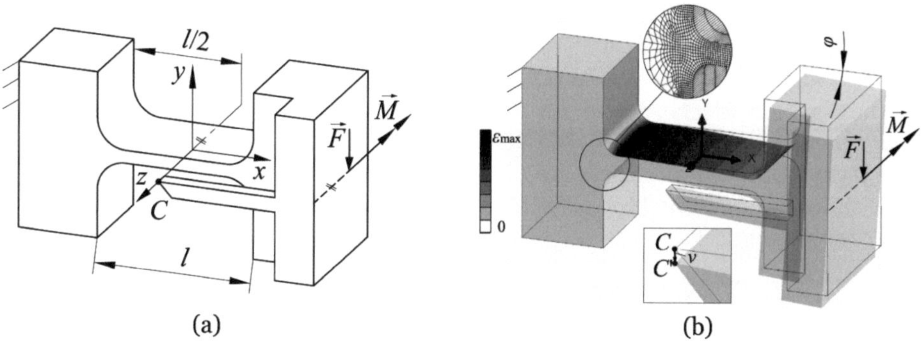

Figure 6.
FEM-based characterization of a flexure hinge: (a) CAD model and (b) FEM model with deformed hinge and mesh details.

characteristic $M(\varphi)$ and $F(\varphi)$ curves can be determined. The maximum deflection angle φ_{max} can be calculated in dependence of the regarded maximum value of the equivalent von Mises strain ε_{max}. Moreover, large deflections are considered for an accurate comparison with the analytical calculations due to the nonlinear beam theory. Other assumptions are a linear material behavior and a comparable and fine discretization of the hinge, especially in areas of the minimum hinge height.

For the determination of the rotational precision, an additional part is added onto the CAD model according to the often used and chosen fixed center approach [36]. Based on guiding the center point C with a constant length $l/2$, the distance between C and C' defines the rotational axis shift v.

3.2 Design graphs

Among the four investigated flexure hinge contours (cf. **Figure 4**), the power function contour allows the modeling of a wide spectrum of different notch hinges. Depending on the exponent n and the hinge dimensions, arbitrary complex curves can be realized, or nearly any elementary geometry can be approximated. For a given deflection angle φ, it has been shown that 16th-order polynomial contours lead to low stress or strain values comparable to those of corner-filleted contours [13]. Furthermore, it was found that fourth-order polynomial contours allow for both a precise rotation and large rotation angles in general [24]. In addition, it is quite possible to realize required hinge properties by arbitrarily varying n.

Based on a geometrically nonlinear FEM simulation using a given displacement at the free end, design graphs for power function-shaped flexure hinges with typical dimension have been created (see **Figure 7**) (cf. [24]). Thus, the bending moment M can be easily determined depending on the rotation angle φ and n. Analogously, the transverse force F results in good approximation by considering the moment M divided by the half-length $L/2$ (cf. **Table 4**) [33]. Furthermore, a minimum required exponent n can be read out depending on φ and the admissible material strain ε_{adm}. The determination of n is also possible with an odd or rational number.

3.3 Nonlinear analytical calculation

As long as the dimensions of the cross section are small compared to the rod length, the nonlinear theory for large deflections of curved rodlike structures is sufficient to describe the deformation behavior of compliant systems [2]. Hence, the analytical investigations are based on the well-known Euler-Bernoulli's theorem for a static problem of a slender structure with an assumed axial inextensible line. The additional assumption is made that Saint-Venant's principle and Hooke's law apply. If a flexure hinge is modeled together with adjacent deformable link segments as a bent rod with a variable height, this theory is assumed to be suitable, too. Further specific effects relevant for notch flexure hinges have to be expected especially for very thin hinges [43], but they are neglected here with regard to general design guidelines. Among them are shear deformation [44], stress concentration [45], anticlastic bending [46], or manufacturing imperfections [47].

For the analytical calculation, a stationary coordinate system $\{\xi, \eta, \zeta\}$ is considered with the origin 0 at the fixed end (cf. **Figure 5b**). The arc length s is used to characterize the neutral axis in the deflected state. The bent hinge undergoes a displacement $u_\xi(s)$ and $u_\eta(s)$ for each coordinate along s, which lead to the deflection angle $\theta(s)$. The curvature $\kappa(s)$ is the gradient of $\theta(s)$. Hence, four nonlinear differential equations are used to model a flexure hinge in the deflected state:

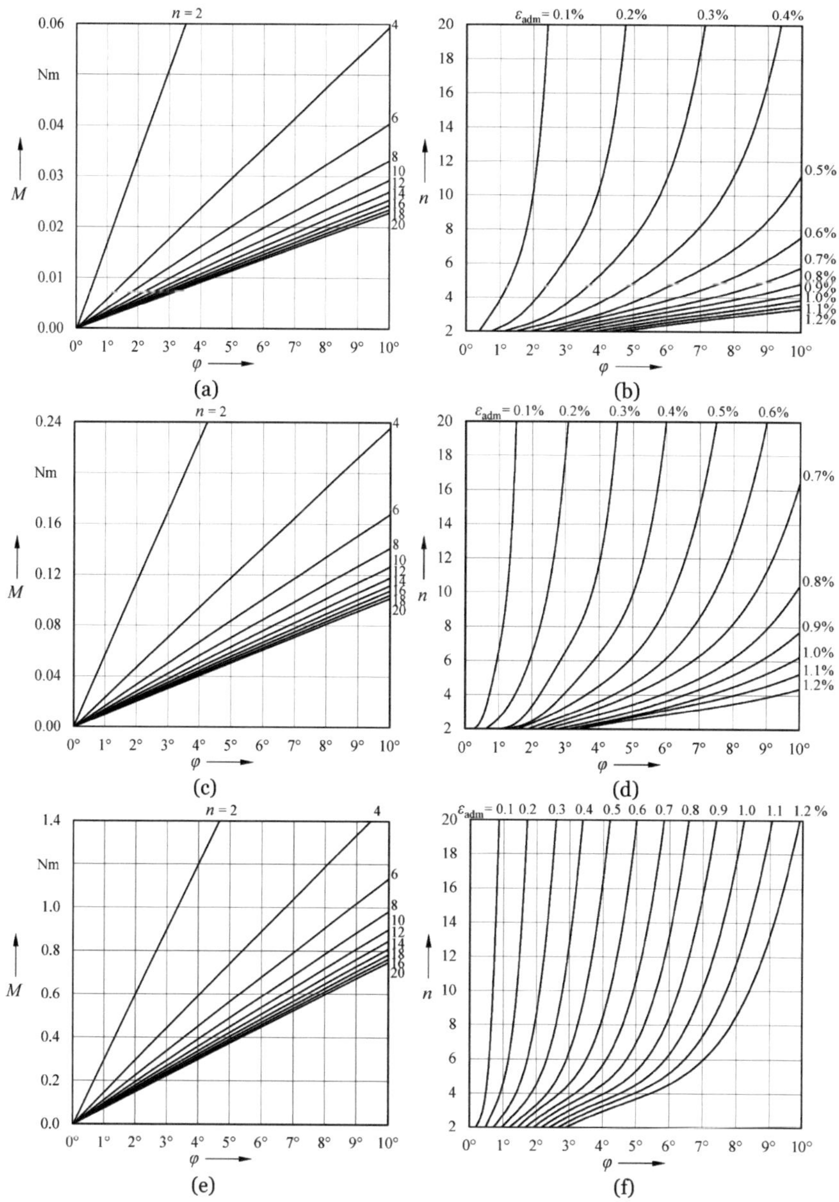

Figure 7.
FEM-based design graphs to determine the bending moment M *(depending on the rotation angle* φ *and the exponent* n *of each power function flexure hinge) or to determine the minimum exponent* n *(depending on* φ *and the admissible material strain* ε_{adm}*), created for different hinge heights and* $\beta_L = 3$, $\beta_l = 1$, $\beta_w = 0.6$ *(cf. [24]): (a)* M *for* $\beta_h = 0.03$*, (b)* n *for* $\beta_h = 0.03$*, (c)* M *for* $\beta_h = 0.05$*, (d)* n *for* $\beta_h = 0.05$*, (e)* M *for* $\beta_h = 0.1$*, and (f)* n *for* $\beta_h = 0.1$*.*

$$\frac{dM_\zeta}{ds} + F\cos\theta = 0, \tag{3}$$

$$\frac{d\theta}{ds} - \kappa = 0, \text{ with } \kappa = \frac{M_\zeta}{E\,I_\zeta}, \tag{4}$$

$$\frac{du_\xi}{ds} - \cos\theta + 1 = 0, \tag{5}$$

$$\frac{du_\eta}{ds} - \sin\theta = 0. \tag{6}$$

The initial curvature is zero here, because a fully symmetric flexure hinge is considered. A numerical solution is done for the system of differential equations with the subsequent boundary conditions for a bending moment at the loaded side:

$$M_\zeta(L) = M, \theta(0) = 0, u_\xi(0) = 0, u_\eta(0) = 0, \tag{7}$$

and with the following conditions for a transverse force load.

$$M_\zeta(L) = 0, \theta(0) = 0, u_\xi(0) = 0, u_\eta(0) = 0. \tag{8}$$

The boundary value problem is solved numerically with MATLAB [14]. At the end of this procedure, all four parameters κ, θ, u_ξ, and u_η are obtained for each point s along the deformed neutral axis, and further results for the considered hinge properties can be estimated. The bending stress σ is analyzed after linear beam theory to characterize the maximum stress of the entire flexure hinge for a given deflection angle. According to Hooke's law, stress and strain are linearly dependent on each other by E, wherefore the bending strain ε is considered. The maximum absolute value of the strain always occurs at the outer fiber for the maximum η-coordinate, which corresponds with the contour height function of the hinge. In addition, the absolute value of the rotational axis shift v, based on the fixed center approach, is put together from the axis shift of point C in ξ and η-direction [14].

3.4 Design equations

To provide closed-form equations which can be used for the simplified flexure hinge synthesis regarding all three rotational properties, six design equations have been developed for both load cases based on the analytical characterization due to the described nonlinear theory (see **Table 3**). SI units must be used for all parameters. The load acts close to the hinge center at $L = 2H$ in this case, while only the elastic properties are almost independent from the value of this distance for the bending moment load [33]. The equations are accurately valid for a rotation angle up to $\varphi = 5°$. The calculation of results for larger angles is nevertheless possible.

With regard to an accelerated and unified synthesis of compliant mechanisms, the general design equations are concise and thus advantageous. With only two coefficients, their structural form is simple, contour-independent, and, with respect to the maximum hinge height or link height H as scaling factor, dimensionless.

The further necessary contour-specific coefficients of all six design equations are given in **Table 4** for the four regarded hinge contours and an appropriate parameter range of the hinge dimensions, the hinge length ratio β_l ($0.5 \le \beta_l \le 1.5$) and the hinge height ratio β_h ($0.03 \le \beta_h \le 0.1$). The coefficients of the used power functions are determined with MATLAB based on a fitting procedure in order to attain the smallest maximum error over all calculated result points [33].

Property	Bending moment	Transverse force
Bending stiffness	$\frac{M}{\varphi} = k_{M1}\, E\, \beta_w\, {\beta_l}^{(-k_{M2})} {\beta_h}^{(2+k_{M2})} H^3$ (9)	$\frac{F}{\varphi} = k_{F1}\, E\, \beta_w\, {\beta_l}^{(-k_{F2})} {\beta_h}^{(2+k_{F2})} H^2$ (10)
Maximum angular deflection	$\lvert\varphi_{\max}\rvert = \frac{\varepsilon_{\text{adm}}}{6\, k_{M1}} \left(\frac{\beta_l}{\beta_h}\right)^{k_{M2}}$ (11)	$\lvert\varphi_{\max}\rvert = \frac{\varepsilon_{\text{adm}}}{12\, (1-k_{\text{crit}}) k_{F1}} \left(\frac{\beta_l}{\beta_h}\right)^{k_{F2}}$ (12)
Rotational precision/axis shift	$\frac{v}{\varphi^2} = k_{vM1}\, {\beta_l}^{k_{vM2}}\, {\beta_h}^{(1-k_{vM2})} H$ (13)	$\frac{v}{\varphi} = k_{vF1}\, {\beta_l}^{k_{vF2}}\, {\beta_h}^{(2-k_{vF2})} H$ (14)

Table 3.
Contour-independent closed-form design equations based on analytical modeling.

Hinge contour	k_{M1} $[10^{-3}]$	k_{M2}	k_{F1} $[10^{-2}]$	k_{F2}	k_{crit}	$k_{\upsilon M1}$ $[10^{-3}]$	$k_{\upsilon M2}$	$k_{\upsilon F1}$ $[10^{-2}]$	$k_{\upsilon F2}$
Circular 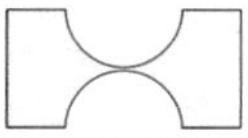	107.90	0.52	10.55	0.51	0.5	99.85	0.52	19.12	0.94
Corner-filleted 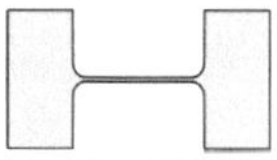	83.95	0.96	8.41	0.96	$0.5–0.2\beta_l$	85.76	0.95	9.20	1.89
Elliptical 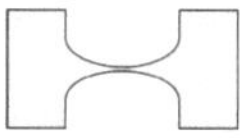	82.50	0.54	8.27	0.54	0.5	114.35	0.57	18.21	1.14
Power function (fourth order) 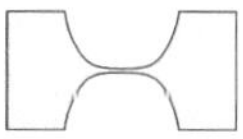	112.07	0.74	11.22	0.74	$0.4\beta_l^{(-0.081)}$ $\beta_h^{(-0.048)}$	71.32	0.73	6.24	1.47

Table 4.
*Load and contour-specific coefficients for the design equations in **Table** 3.*

The relative discrepancy errors between the design equation results and the analytical results, a comparison with FEM results, as well as coefficients for further power function contours are mentioned in [48]. According to the theory, the accuracy of the results is nearly independent of the parameter range for the hinge width β_w. The maximum strain occurs contour-independently in the hinge center for a bending moment. An additional factor k_{crit} has become necessary to consider the deviation of the critical strain location from the center for a transverse force load, especially for corner-filleted and power function hinges (cf. Section 3.7).

3.5 Design tool detasFLEX

Moreover, computational design tools may prove useful for the comprehensive analysis and synthesis of various notch flexure hinges, such as the developed tool detasFLEX [14], which is also based on the described nonlinear modeling approach (cf. Section 3.3). The graphical user interface (GUI) is shown in **Figure 8**.

The design tool was created with MATLAB as a stand-alone software application which only requires the license-free runtime environment. Four flexure hinge contours are considered, the circular, corner-filleted, elliptical, and power function-based contours (cf. **Figure 4**). Various geometric and material parameters may be realized to allow for a broad usability in different cases. The calculation is possible for a bending moment and a transverse force as well as both loads combined for different lengths of each hinge side. The computation of results is further possible for all three load cases with a given load or a given rotation angle up to 45°. A wide range of result parameters may be computed, and the most important hinge performance properties like the deformed neutral axis, the bending stiffness, the rotational precision, and the elastic strain distribution are illustrated in the form of diagrams. Additionally, a preview of the exact hinge geometry with the instant visualization of input changes is implemented. Also, values for the angle or load, axis shift, strain distribution, maximum strain, and maximum possible rotation angle are calculated. Using a corner-filleted hinge, for example, the deviation of the bending stiffness between the FEM and design tool results is in the range of 0.1–9.4% for a given rotation angle of 10° [14].

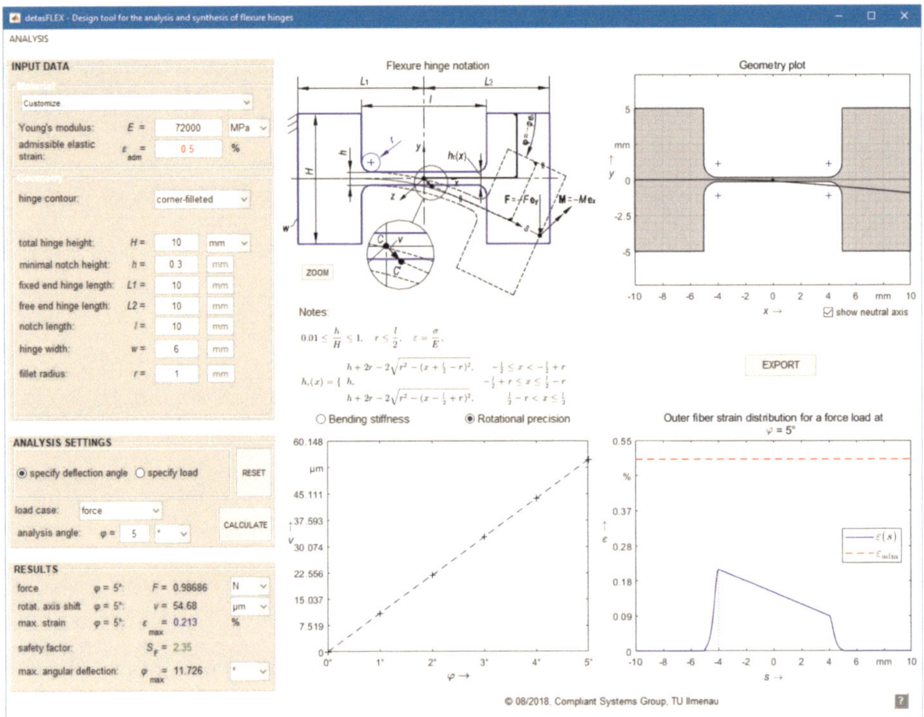

Figure 8.
Graphical user interface of the computational design tool detasFLEX, shown using the example of a corner-filleted flexure hinge.

DetasFLEX enables a wide variety of different geometry, material and contour selections, as well as multiple analysis criteria and settings so that numerous notch flexure hinges for diverse tasks may be accurately analyzed within a few seconds. Thus, each hinge in a compliant mechanism can be designed purposefully and individually. Based on this, the PC-based synthesis is generally possible, too.

3.6 Comparison of results and usability

The different methods for modeling the elasto-kinematic flexure hinge properties described above are compared in **Table 5** using the example of a power function-shaped hinge of the fourth order. The design tool results and analytical modeling are mentioned together due to the equality of the values. It is obvious that

Method	Bending moment				Transverse force			
	$\varphi = 5°$			$\varepsilon_{adm} = 0.5\%$	$\varphi = 5°$			$\varepsilon_{adm} = 0.5\%$
	M [Nm]	v [µm]	ε_{max} [%]	φ_{max} [°]	F [N]	v [µm]	ε_{max} [%]	φ_{max} [°]
FEM	0.0294	2.190	0.414	6.039	2.946	9.980	0.439	5.695
Design graph	0.029	—	0.43	5.3	2.9	—	0.43	5.3
Design equation	0.0284	2.107	0.438	5.707	2.842	8.490	0.459	5.403
Design tool/ analytic	0.0277	2.226	0.428	5.839	2.785	9.459	0.450	5.562

Table 5.
Comparison of results for the method-dependent elasto-kinematic properties using the example of a power function hinge of the fourth order ($\beta_L = 2$, $\beta_l = 1$, $\beta_h = 0.03$, $\beta_w = 0.6$).

the suggested design guidelines and tools allow the accurate and simplified determination or calculation of the deformation, stress/strain, and motion behavior with respect to the assumptions and geometric restrictions.

Regarding the usability, the design tool provides the most comprehensive support for the modeling and design of various notch flexure hinges (see **Table 6**).

Additionally, the design equations are also easy to use for the four regarded hinge contours. Furthermore, it becomes obvious that the determination method influences the possible values for the hinge dimensions and the power function exponent n as well as valid ranges of the deflection angle. Due to compactness, in Section 4.3, n is exemplarily determined based on the design graph approach.

In conclusion, all three design aids can be used for the accelerated contour-specific quasi-static analysis of the elasto-kinematic properties of notch flexure hinges with no need for further iterative and time-consuming simulations. Moreover, the guidelines and tools may be used for the systematic angle-dependent synthesis of compliant mechanisms with differently optimized flexure hinges (cf. Section 4).

3.7 Influence of the contour on the elasto-kinematic hinge properties

Independent of the selected method, the influence of the flexure hinge contour on the elasto-kinematic hinge properties can be generalized, especially for thin hinges. In **Figure 9**, the analytical results are exemplarily presented for a force load.

The load-angle behavior is almost linear for the regarded angular deflection up to 5°. The following order can be concluded from the lowest to the highest stiffness: the corner-filleted, power function, elliptical, and circular contour (**Figure 9a**).

Because the maximum strain value limits the deflection, the maximum rotation angle of a flexure hinge is always possible with a corner-filleted contour, while a circular contour leads to the lowest possible angles (**Figure 9c**). Furthermore, the asymmetric strain distribution due to the transverse force load is obvious, especially for a corner-filleted contour (**Figure 9d**). Due to the notch effect, the strain is concentrated in the hinge center for a circular and elliptical contour, while the other contours lead to a more even strain distribution along the hinge length.

Furthermore, the hinge contour has a strong influence on the axis shift, which can be in the range of some micrometers up to submillimeters in dependence of β_l and especially β_h. With regard to a high rotational precision or a small axis shift, the following order is existing for thin hinges: the circular contour, elliptical contour or power function contour to the same extend, and corner-filleted contour (**Figure 9b**).

Thus, the power function contour of the fourth order simultaneously provides a large angular deflection and a high rotational precision. The influence of the basic hinge dimensions is further investigated in [33]. An influence of β_L is to be expected, too.

4. Modeling and design of compliant mechanisms

In this section, the synthesis method presented in Section 2.1 is applied to a path-generating mechanism to explore the angle-based approach of the optimal design with individually shaped flexure hinges in one single compliant mechanism using power functions. Therefore, a symmetric four-bar Roberts mechanism with four hinges, realizing the guidance of the coupler point P on an approximated rectilinear path (cf. [19–22]), is investigated. The rigid-body model and the compliant mechanism are shown in **Figure 1** in the initial and deflected positions.

Method, related reference	Hinge contours				Domain and value of n	Hinge dimensions β_L, β_l, β_h, and β_w	Range of φ	Result criteria	Modeling effort/ computation time
	Circular	Corner-filleted	Elliptical	Power function					
FEM, nonlinear, e.g., [24]	x	x	x	x	Arbitrary	Arbitrary	Arbitrary	Arbitrary	Great/ high
Analytical, nonlinear, e.g., [2]	x	x	x	x	Arbitrary	Arbitrary	Arbitrary	Arbitrary	Great/ low
Design graph [24]				x	$2 \leq n \leq 20$	Predefined (three cases for β_h)	$\leq 10°$	$M(\varphi)$, $(F(\varphi))$, $\varphi_{\max}$, $\varepsilon_{\max}$	None
Design Eq. [33, 48]	x	x	x	x	2, 3, 4, 8, 16	Constrained	$\leq 5°$	$M(\varphi)$, $F(\varphi)$, $v(\varphi)$, $\varphi_{\max}$, $\varepsilon_{\max}$	None
Design tool [14]	x	x	x	x	$1.1 \leq n \leq 50$	Slightly constrained	$\leq 45°$	$M(\varphi)$, $F(\varphi)$, $v(\varphi)$, $\varphi_{\max}$, $\varepsilon_{\max}$, $\varepsilon(s)$	Little/low

Table 6.
Comparison of usability of the presented methods, guidelines, and approaches for the modeling and design of notch flexure hinges.

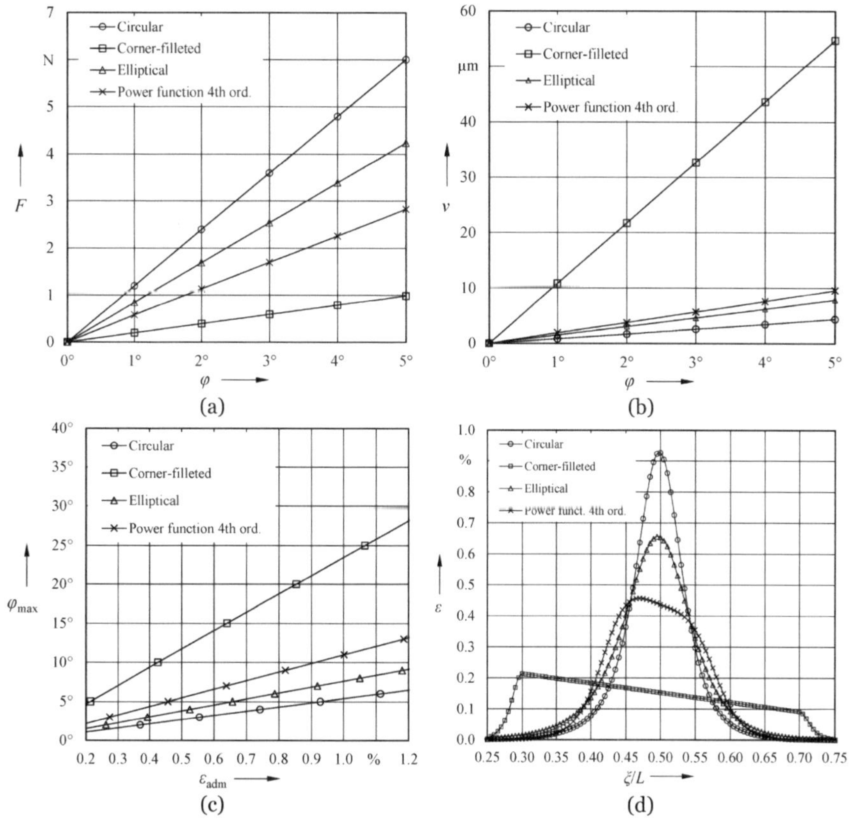

Figure 9.
Analytical results for the influence of the flexure hinge contour on the hinge properties ($\beta_L = 2$, $\beta_l = 1$, $\beta_h = 0.03$, $\beta_w = 0.6$, force load): (a) bending stiffness, (b) axis shift, (c) maximum angular deflection, and (d) outer fiber strain distribution for $\varphi = 5°$.

The link lengths are suitably chosen as $\overline{A_0A} = \overline{B_0B} = 66.6$ mm, $\overline{AB} = 56.6$ mm, $\overline{A_0B_0} = 165.71$ mm, and $\overline{AP} = \overline{BP} = 73.6$ mm, according to [24]. Furthermore, the mechanism position with the replacement angle $\gamma = 35°$ is used. Thus, applying the input displacement $s = u_{xP}^* = u_{xP} = 10$ mm in point P, the relevant straight-line deviation for the rigid-body model results as $u_{yP}^* = -24.7$ μm.

4.1 Synthesis method based on individually shaped flexure hinges

A compliant mechanism with individually shaped power function flexure hinges is synthesized according to the synthesis method based on the relative rotation angles in the rigid-body model (cf. Section 2.1) exemplarily using the design graph approach (cf. Section 3.2). The resulting compliant mechanism is shown in **Figure 10d**. Furthermore, the mechanism properties are compared with three compliant mechanisms using identical hinges designed with circular, corner-filleted, or power function contours of the fourth order (see **Figure 10a–c**).

Following the rigid-body replacement approach, the flexure hinge centers are designed identical to the revolute joints. Next, suitable flexure hinge orientations are chosen with respect to the link orientations of the crank and the coupler (cf. Section 4.3). The main link parameters are specified as $H = 10$ mm and $\beta_w = 0.6$, while the same aluminum material as in Section 3 is used ($E = 72$ GPa, $\varepsilon_{adm} = 0.5\%$). Furthermore, comparable short and thin hinges are used with the hinge length ratio $\beta_l = 1$ and the height ratio $\beta_h = 0.03$ because they are especially suitable [24].

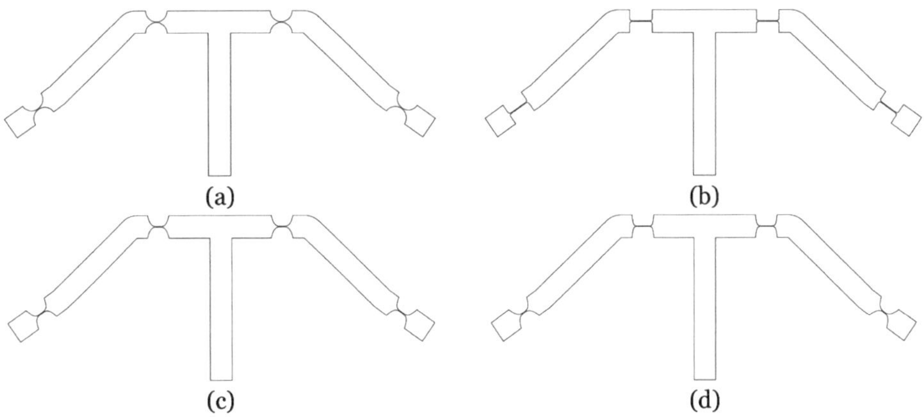

Figure 10.
Designs of the compliant Roberts mechanism (β_l = 1, β_h = 0.03) with (a) circular hinges (R = H/2), (b) corner-filleted hinges (r = 0.1 l), (c) identical power function hinges (n = 4), and (d) different power function hinges (n_{Ao} = 3, n_A = 12, n_B = 10, n_{Bo} = 3).

Based on the relative rigid-body-based rotation angles $\varphi_{A0}{}^* = 3.8°$, $\varphi_A{}^* = 10.4°$, $\varphi_B{}^* = 9.5°$, and $\varphi_{B0}{}^* = 3.0°$, which result through a simple kinematic analysis, the power function exponents can be determined for the assumption $\varphi = \varphi^*$. According to the designed graph in **Figure 7b**, the exponents result in $n_{A0} = 3$, $n_A = 12$, $n_B = 10$, and $n_{B0} = 3$. The exponents are exemplarily determined as even numbers, while rational exponents are also possible for a more specific design.

4.2 Nonlinear FEM simulation

For the quasi-static structural and geometrically nonlinear FEM simulation of the compliant Roberts mechanisms, the same settings as for a separate hinge are used (cf. Section 3.1). The results for the motion path of the coupler point *P* are shown in **Figure 11** for a given *x*-displacement in dependence of the used flexure hinge contours and additionally for the rigid-body model. Regarding a consistent modeling, the coordinate system is defined at the fixed support in the following.

The results for the path deviation compared with the rigid-body model confirm the impact of the synthesis approach for the mechanism with different power

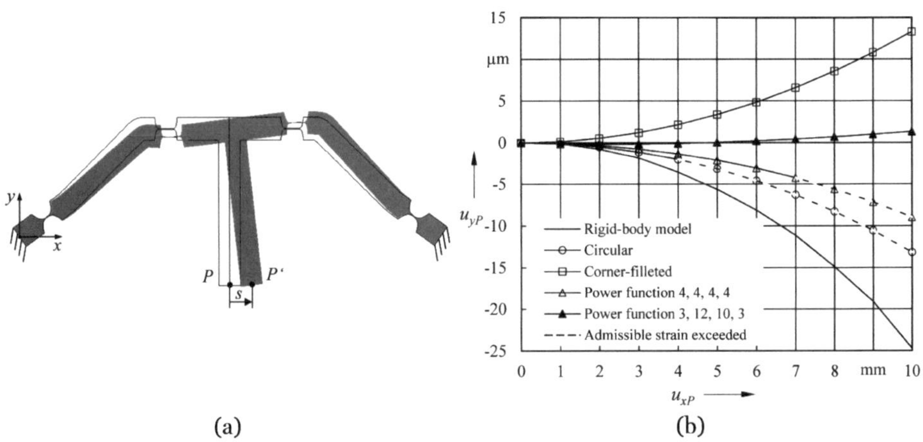

Figure 11.
FEM results for the motion behavior of the Roberts mechanisms (β_l = 1, β_h = 0.03): (a) model and deformed state for input s = u_{xP} = 10 mm and (b) straight-line deviation of P; the curves are drawn in dashed lines from the input displacement at which ε_{adm} is exceeded.

function contours regarding a higher path precision (compared to identical corner-filleted contours) and the possible required large stroke (compared to identical semicircular and power function contours with $n = 4$). Furthermore, analyzing the straight-line deviation, it becomes obvious that a more precise rectilinear motion can be realized using the compliant mechanism with individually shaped flexure hinges.

4.3 Nonlinear analytical calculation

The analytical modeling of the compliant mechanisms is also based on the nonlinear theory for large deflections of rodlike structures described in Section 3.3. To consider the coupler point P, a branched mechanism has to be modeled, and the rod is split into three sections in K, each with its own rod axis s_1–s_3 (see **Figure 12**). A force-driven analysis is implemented, while the input force is increased until the desired displacement $u_{xP} = 10$ mm is reached. The straight-line deviation, the maximum strain, and the necessary deflection force are determined, too.

From investigations on separate hinges [49] and flexure hinge-based compliant mechanisms [50, 51], it is known that the flexure hinge orientation strongly influences the elasto-kinematic properties of compliant mechanisms. Therefore, a study of the Roberts mechanism is done, while the hinges A_0 and B_0 and A and B are modeled equally mirrored (see **Figure 13**). From the results, the suitable

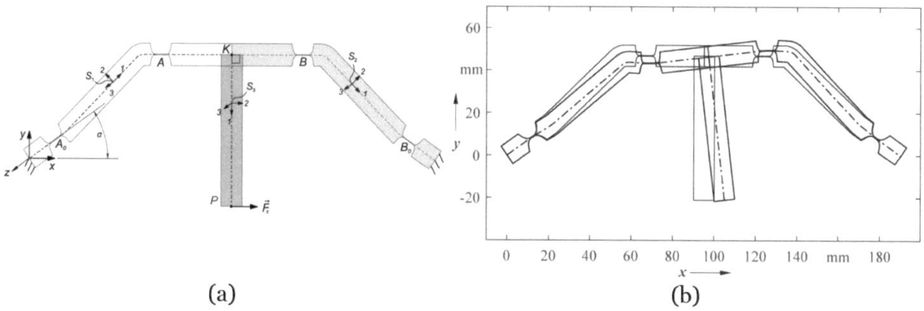

Figure 12.
Analytical modeling of the Roberts mechanism ($\beta_l = 1$, $\beta_h = 0.03$): (a) branched mechanism with point P split up in three individual rods connected at point K, (b) resulting from MATLAB plot of initial and deflected state for $u_{xP} = 10$ mm ($n_{A_0} = 3$, $n_A = 12$, $n_B = 10$, $n_{B_0} = 3$).

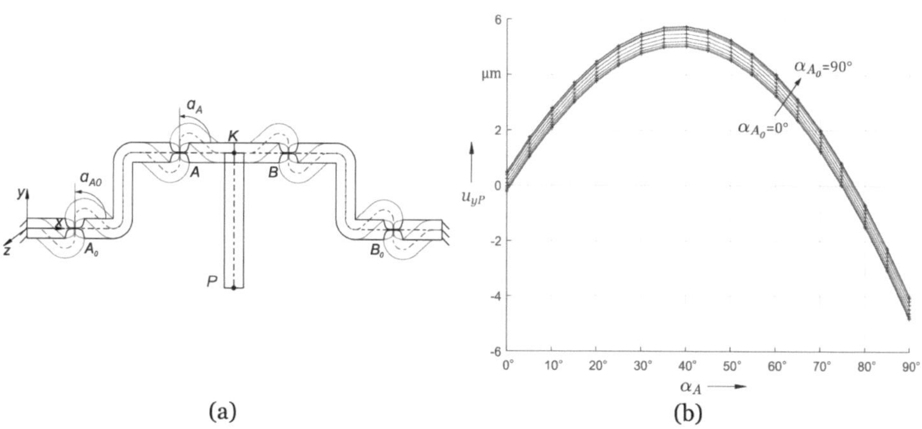

Figure 13.
Analytical results for the influence of the flexure hinge orientation on the straight-line deviation of a mechanism with power function contours ($\beta_l = 1$, $\beta_h = 0.03$, $u_{xP} = 10$ mm): (a) model with two variable orientation angles from 0° to 90° and (b) deviation of P.

orientations α_{A0} = 35° and α_A = 0° can be concluded to realize the smallest straight-line deviation.

4.4 Comparison of results

The FEM results and analytical results for the four investigated compliant Roberts mechanisms are in a very good correlation (see **Table** 7).

Generally, all four compliant mechanisms exhibit a very small straight-line deviation in the low micrometer range. With respect to the path deviation compared to rigid-body model, the values differ from the straight-line deviations. However, as for the separate hinge (cf. **Figure 9b**), the mechanism with circular contours provides the smallest path deviation. With regard to the maximum admissible strain, the desired stroke cannot be realized when using identical circular or power function hinges of the fourth order (cf. **Figure 11b**). In contrast, the full stroke is possible when using the corner-filleted hinges and, as expected, also with the synthesized mechanism with individually shaped hinges. Furthermore, the input force varies considerably, and, thus, a required stiffness can be achieved, too.

Hence, the result method independently confirms the practicability and impact of the angle-based synthesis method for different hinges in one mechanism. Moreover, the presented nonlinear analytical approach is suitable to accurately model the elasto-kinematic properties of planar flexure hinge-based compliant mechanisms under consideration of the specific hinge contour without simulations.

4.5 Geometric scaling approach

The influence of the scale on the deformation and motion behavior is a further relevant aspect regarding the similitude of mechanisms [52]. Based on investigations of a separate flexure hinge and a compliant parallel linkage [53], the uniform geometric scaling may also be a suitable synthesis approach for compliant mechanisms if the change ratios of the elasto-kinematic properties are known.

Here, uniform geometric scaling is understood as a linear variation of all geometric length parameters with the scale factor of the value g, while the initial dimension H = 10 mm is considered for g = 1. The geometric scaling approach is exemplified for the Roberts mechanism with different power function hinges based

Hinge contours	Method	Straight-line deviation u_{yP} [µm]	Path deviation $\lvert u_{yP} - u_{yP}^* \rvert$ [µm]	Input force F_x [N]	Maximum strain ε_{max} [%]
Identical circular, $R = H/2$ (**Figure 10a**)	FEM	−13.20	11.53	4.93	1.84
	Analytical	−13.72	11.02	4.61	1.88
Identical corner-filleted, $r = 0.1\,l$ (**Figure 10b**)	FEM	13.20	37.93	0.80	0.36
	Analytical	13.65	38.38	0.78	0.33
Identical power function, $n = 4$ (**Figure 10c**)	FEM	−8.99	15.74	2.13	0.85
	Analytical	−9.23	15.50	2.16	0.89
Different power function (**Figure 10d**)	FEM	1.29	26.02	1.30	0.46
	Analytical	0.86	25.59	1.25	0.47

Table 7.
Comparison of FEM and analytical results for the elasto-kinematic properties of the Roberts mechanisms with identical common hinge contours and with different power function contours (β_l = 1, β_h = 0.03, β_w = 0.6, with H = *10 mm and input* u_{xP} = *10 mm).*

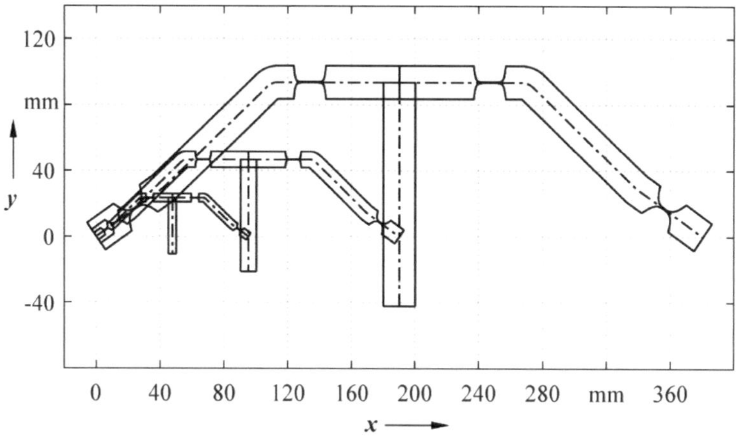

Figure 14.
Geometric scaling of the compliant Roberts mechanism shown for the factors g = 0.5, g = 1, *and* g = 2 *(mechanism with different power function hinges,* $\beta_l = 1$, $\beta_h = 0.03$*).*

Scaling factor	Stroke u_{xP} [mm]	Straight-line deviat. u_{yP}^* [µm]	Straight-line deviat. u_{yP} [µm]	Path deviation $\|u_{yP} - u_{yP^*}\|$ [µm]	Input force F_x [N]	Max. strain ε_{max} [%]	Angle φ_A [°]
$g = 0.5$	5	−10.78	0.431	11.21	0.123	0.468	10.33
$g = 1$	10	−24.73	0.862	25.59	1.249	0.468	10.33
$g = 2$	20	−49.74	1.725	51.46	4.997	0.468	10.33
$g = 10$	100	−249.91	8.623	258.53	124.930	0.468	10.33

Table 8.
Analytical results for the influence of geometric scaling on the elasto-kinematic properties for the mechanism with different power function hinges ($\beta_l = 1$, $\beta_h = 0.03$, $\beta_w = 0.6$*).*

on analytical calculations. Therefore, different scaling factors are regarded (see **Figure 14**). The results are mentioned in **Table 8**, while the input stroke u_{xP} is scaled as well.

Based on the results, geometric scaling is an appropriate approach for the accelerated synthesis through the adjustment of an initially designed or used compliant mechanism with known elasto-kinematic properties to each required scale of the new application through the use of the property change ratios concluded in **Table 9**. The ratio is defined as property value for $g \neq 1$ related to property value for $g = 1$. Therefore, performing a new calculation is not necessary anymore. The results are independent of the hinge contour, while a nonuniform scaling is possible, too [53]. Furthermore, this approach can be used to significantly increase the stroke by increasing the mechanism size or to reduce the straight-line deviation by miniaturization because strain values and angles are independent of the scale.

Property	Property change ratio
Maximum strain	1
Angular deflection	1
Input displacement, motion path coordinates, path deviations	g
Input/deflection force	g^2

Table 9.
Influence of uniform geometric scaling with the factor g *on the change ratios of the elasto-kinematic properties of planar flexure hinge-based compliant mechanisms.*

5. Conclusions

Flexure hinge-based compliant mechanisms offer a high-precise and large-stroke guidance motion with straight-line or path deviations in the single-micrometer range if they are purposefully designed. It is shown that the synthesis of a compliant mechanism with individually shaped flexure hinges based on the relative rotation angles in the rigid-body model is a suitable and general synthesis method which is easy to use without the need of numerical calculations, FEM simulations, or a multi-criterial optimization (cf. [25, 35]). Therefore, this chapter provides a survey of several approaches, guidelines, and aids for the accurate and comprehensive design of notch flexure hinges using various hinge contours, while power function contours are particularly suitable. The use of design graphs, design equations, a computational design tool, or a geometric scaling approach is briefly presented. The results are verified by analytical calculations and FEM simulations, and also, not mentioned, by experimental investigations (e.g., [3, 24, 33]). Moreover, especially the used nonlinear analytical approach has a great potential for the future work, for example, the implementation of a GUI for the compliant mechanism synthesis.

Acknowledgements

We acknowledge support for the research by the DFG (Grant no. ZE 714/10-2). We further acknowledge support for the Publishing Process Charge by the Thuringian Ministry for Economic Affairs, Science and Digital Society and the Open Access Publication Fund of the Technische Universität Ilmenau.

Conflict of interest

We declare that we have no conflict of interest.

Author details

Sebastian Linß*, Stefan Henning and Lena Zentner
Compliant Systems Group, Technische Universität Ilmenau, Ilmenau, Germany

*Address all correspondence to: sebastian.linss@tu-ilmenau.de

References

[1] Howell LL, Magleby SP, Olsen BM. Handbook of Compliant Mechanisms. Chichester: Wiley; 2013. 324 p

[2] Zentner L. Nachgiebige Mechanismen. München: De Gruyter Oldenbourg; 2014. 133 p

[3] Gräser P, Linß S, Harfensteller F, Zentner L, Theska R. Large stroke ultra-precision planar stage based on compliant mechanisms with polynomial flexure hinge design. In: Proceedings of the 17th Euspen; Hannover, Germany. 2017. pp. 207-208

[4] Teo TJ, Yang G, Chen I-M. A large deflection and high payload flexure-based parallel manipulator for UV nanoimprint lithography: Part I. Modeling and analyses. Precision Engineering. 2014;**38**(4):861-871. DOI: 10.1016/j.precisioneng.2014.05.003

[5] Beroz J, Awtar S, Bedewy M, Tawfick S, Hart AJ. Compliant microgripper with parallel straight-line jaw trajectory for nanostructure manipulation. In: Proceedings of the 26th Annual Meeting of the ASPE; Denver, USA. 2011

[6] Darnieder M, Pabst M, Wenig R, Zentner L, Theska R, Fröhlich T. Static behavior of weighing cells. Journal of Sensors and Sensor Systems. 2018;**7**(2): 587-600. DOI: 10.5194/jsss-7-587-2018

[7] Lobontiu N. Compliant Mechanisms: Design of Flexure Hinges. Boca Raton, Fla: CRC Press; 2003. 447 p

[8] Pavlovic NT, Pavlovic ND. Compliant mechanism design for realizing of axial link translation. Mechanism and Machine Theory. 2009; **44**(5):1082-1091. DOI: 10.1016/j. mechmachtheory.2008.05.005

[9] Xu Q. Design and Implementation of Large-Range Compliant Micropositioning Systems. Singapore: John Wiley & Sons Singapore Pte. Ltd; 2016. 273 p

[10] Henein S, Spanoudakis P, Droz S, Myklebust LI, Onillon E. Flexure pivot for aerospace mechanisms. In: Proceedings of the 10th European Space Mechanisms and Tribology Symposium; San Sebastian, Spain. 2003

[11] Cosandier F, Eichenberger A, Baumann H, Jeckelmann B, Bonny M, Chatagny V, et al. Development and integration of high straightness flexure guiding mechanisms dedicated to the METAS watt balance mark II. Metrologia. 2014;**51**(2):88-95. DOI: 10.1088/0026-1394/51/2/S88

[12] Lin C-F, Shih C-J. Multiobjective design optimization of flexure hinges for enhancing the performance of micro-compliant mechanisms. Journal of the Chinese Institute of Engineers. 2005;**28**(6):999-1003. DOI: 10.1080/ 02533839.2005.9671075

[13] Linß S, Erbe T, Zentner L. On polynomial flexure hinges for increased deflection and an approach for simplified manufacturing. In: Proceedings of the 13th World Congress in Mechanism and Machine Science; Guanajuato, Mexico. 2011. A11_512

[14] Henning S, Linß S, Zentner L. detasFLEX—A computational design tool for the analysis of various notch flexure hinges based on non-linear modeling. Mechanical Sciences. 2018;**9**(2):389-404. DOI: 10.5194/ms-9-389-2018

[15] Howell LL, Midha A. A method for the design of compliant mechanisms with small-length flexural pivots. Journal of Mechanical Design. 1994; **116**(1):280-290. DOI: 10.1115/1.2919359

[16] Frecker MI, Ananthasuresh GK, Nishiwaki S, Kota S. Topological

synthesis of compliant mechanisms using multi-criteria optimization. Journal of Mechanical Design. 1997; **119**(2):238-245. DOI: 10.1115/1.2826242

[17] Hopkins JB, Culpepper ML. Synthesis of multi-degree of freedom, parallel flexure system concepts via freedom and constraint topology (FACT)—Part I: Principles. Precision Engineering. 2010;**34**(2):259-270. DOI: 10.1016/j.precisioneng.2009.06.008

[18] Pavlovic ND, Petkovic D, Pavlovic NT. Optimal selection of the compliant mechanism synthesis method. In: Proceedings of the International Conference Mechanical Engineering in XXI Century; Niš, Serbia. 2010. pp. 247-250

[19] Pavlovic NT, Pavlovic ND. Motion characteristics of the compliant four-bar linkages for rectilinear guiding. Journal of Mechanical Engineering Design. 2003;**6**(1):20-27

[20] Hricko J. Straight-line mechanisms as one building element of small precise robotic devices. Applied Mechanics and Materials. 2014;**613**:96-101. DOI: 10.4028/www.scientific.net/AMM.613.96

[21] Wan S, Xu Q. Design and analysis of a new compliant XY micropositioning stage based on Roberts mechanism. Mechanism and Machine Theory. 2016; **95**:125-139. DOI: 10.1016/j.mechmachtheory.2015.09.003

[22] Li J, Chen G. A general approach for generating kinetostatic models for planar flexure-based compliant mechanisms using matrix representation. Mechanism and Machine Theory. 2018;**129**:131-147. DOI: 10.1016/j.mechmachtheory.2018.07.015

[23] Clark L, Shirinzadeh B, Zhong Y, Tian Y, Zhang D. Design and analysis of a compact flexure-based precision pure rotation stage without actuator redundancy. Mechanism and Machine Theory. 2016;**105**:129-144. DOI: 10.1016/j.mechmachtheory.2016.06.017

[24] Linß S. Ein Beitrag zur geometrischen Gestaltung und Optimierung prismatischer Festkörpergelenke in nachgiebigen Koppelmechanismen [doctoral thesis]. Ilmenau: TU Ilmenau; 2015. URN: urn:nbn:de:gbv:ilm1-2015000283

[25] Linß S, Milojevic A, Pavlovic ND, Zentner L. Synthesis of compliant mechanisms based on goal-oriented design guidelines for prismatic flexure hinges with polynomial contours. In: Proceedings of the 14th World Congress in Mechanism and Machine Science; Taipei, Taiwan. 2015. DOI: 10.6567/IFToMM.14TH.WC.PS10.008

[26] Meng Q. A design method for flexure-based compliant mechanisms on the basis of stiffness and stress characteristics [doctoral thesis]. Bologna: Universität Bologna; 2012. DOI: 10.6092/unibo/amsdottorato/4734

[27] Carbone G, Liang C, Ceccarelli M, Burisch A, Raatz A. Design and simulation of a binary actuated parallel micro-manipulator. In: Proceedings of the 13th World Congress in Mechanism and Machine Science; Guanajuato, Mexico. 2011. A12_332

[28] Wuest W. Blattfedergelenke für Meßgeräte. Feinwerktechnik. 1950; **54**(7):167-170

[29] Jensen BD, Howell LL. The modeling of cross-axis flexural pivots. Mechanism and Machine Theory. 2002; **37**(5):461-476. DOI: 10.1016/S0094-114X(02)00007-1

[30] Bi S, Zhao S, Zhu X. Dimensionless design graphs for three types of annulus-shaped flexure hinges. Precision Engineering. 2010;**34**(3):

659-667. DOI: 10.1016/j.precisioneng.2010.01.002

[31] Paros JM, Weisbord L. How to design flexure hinges. Machine Design. 1965;**25**(11):151-156

[32] Zentner L, Linß S. Compliant Systems – Mechanics of Elastically Deformable Mechanisms, Actuators and Sensors. München: De Gruyter Oldenbourg; 2019. 166 p

[33] Linß S, Schorr P, Zentner L. General design equations for the rotational stiffness, maximal angular deflection and rotational precision of various notch flexure hinges. Mechanical Sciences. 2017;**8**(1):29-49. DOI: 10.5194/ms-8-29-2017

[34] Zhu BL, Zhang XM, Fatikow S. Design of single-axis flexure hinges using continuum topology optimization method. Science in China/E. 2014;**57**(3): 560-567. DOI: 10.1007/s11431-013-5446-4

[35] Gräser P, Linß S, Zentner L, Theska R. Optimization of compliant mechanisms by use of different polynomial flexure hinge contours. In: Proceedings of the 3rd IAK, Interdisciplinary Applications of Kinematics; Lima, Peru. 2018. DOI: 10.1007/978-3-030-16423-2_25

[36] Linß S, Erbe T, Theska R, Zentner L. The influence of asymmetric flexure hinges on the axis of rotation. In: Proceedings of the 56th International Scientific Colloquium; Ilmenau, Germany. 2011. URN: urn:nbn:de:gbv:ilm1-2011iwk-006:6

[37] Zettl B, Szyszkowski W, Zhang WJ. On systematic errors of two-dimensional finite element modeling of right circular planar flexure hinges. Journal of Mechanical Design. 2005; **127**(4):782-787. DOI: 10.1115/1.1898341

[38] Yong YK, Lu T-F, Handley DC. Review of circular flexure hinge design equations and derivation of empirical formulations. Precision Engineering. 2008;**32**(2):63-70. DOI: 10.1016/j.precisioneng.2007.05.002

[39] Valentini PP, Pennestrì E. Elasto-kinematic comparison of flexure hinges undergoing large displacement. Mechanism and Machine Theory. 2017; **110**:50-60. DOI: 10.1016/j.mechmachtheory.2016.12.006

[40] Venanzi S, Giesen P, Parenti-Castelli V. A novel technique for position analysis of planar compliant mechanisms. Mechanism and Machine Theory. 2005;**40**(11):1224-1239. DOI: 10.1016/j.mechmachtheory.2005.01.009

[41] Raatz A. Stoffschlüssige Gelenke aus pseudo-elastischen Formgedächtnislegierungen in Pararellrobotern [doctoral thesis]. Braunschweig: TU Braunschweig; 2006

[42] Zelenika S, Munteanu MG, De Bona F. Optimized flexural hinge shapes for microsystems and high-precision applications. Mechanism and Machine Theory. 2009;**44**(10):1826-1839. DOI: 10.1016/j.mechmachtheory.2009.03.007

[43] Torres Melgarejo MA, Darnieder M, Linß S, Zentner L, Fröhlich T, Theska R. On Modeling the bending stiffness of thin semi-circular flexure hinges for precision applications. Actuators. 2018;**7**(4):86. DOI: 10.3390/act7040086

[44] Tseytlin YM. Notch flexure hinges: An effective theory. The Review of Scientific Instruments. 2002;**73**(9): 3363-3368. DOI: 10.1063/1.1499761

[45] Dirksen F, Lammering R. On mechanical properties of planar flexure hinges of compliant mechanisms. Mechanical Sciences. 2011;**2**:109-117. DOI: 10.5194/ms-2-109-2011

[46] Campanile LF, Hasse A. A simple and effective solution of the elastica problem. The Proceedings of the Institution of Mechanical Engineers, Part C: Journal of Mechanical Engineering Science. 2008;**222**(12): 2513-2516. DOI: 10.1243/09544062JMES1244

[47] Ryu JW, Gweon D-G. Error analysis of a flexure hinge mechanism induced by machining imperfection. Precision Engineering. 1997;**21**(2/3):83-89. DOI: 10.1016/S0141-6359(97)00059-7

[48] Linß S, Schorr P, Henning S, Zentner L. Contour-independent design equations for the calculation of the rotational properties of commonly used and polynomial flexure hinges. In: Proceedings of the 59th International Scientific Colloquium; Ilmenau, Germany. 2017. URN: urn:nbn:de:gbv:ilm1-2017iwk-001:5

[49] Schorr P, Linß S, Zentner L, Zimmermann K. Influence of the orientation of flexure hinges on the elasto-kinematic properties. In: Tagungsband Vierte IFToMM D-A-CH Konferenz 2018; Lausanne, Switzerland. 2018. DOI: 10.17185/duepublico/45330

[50] Gräser P, Linß S, Zentner L, Theska R. On the influence of the flexure hinge orientation in planar compliant mechanisms for ultra-precision applications. In: Proceedings of the 59th International Scientific Colloquium; Ilmenau, Germany. 2017. URN: urn:nbn:de:gbv:ilm1-2017iwk-090:9

[51] Hao G, Yu J, Liu Y. Compliance synthesis of a class of planar compliant parallelogram mechanisms using the position space concept. In: Proceedings of the 4th ReMAR Conference; Delft, The Netherlends. 2018. DOI: 10.1109/REMAR.2018.8449882

[52] Laudahn S, Sviberg M, Wiesenfeld L, Haberl F, Haidl J, Abdul-Sater K, et al. Similitude of scaled and full scale linkages. In: Proceedings of the 7th European Conference on Mechanism Science: EuCoMeS; Aachen, Germany. 2018. pp. 256-264. DOI: 10.1007/978-3-319-98020-1_30

[53] Linß S, Gräser P, Räder T, Henning S, Theska R, Zentner L. Influence of geometric scaling on the elasto-kinematic properties of flexure hinges and compliant mechanisms. Mechanism and Machine Theory. 2018;**125**(C): 220-239. DOI: 10.1016/j.mechmachtheory.2018.03.008

Chapter 4

Kinetostatic Nonlinear Stiffness Characteristic Generation Using the Kinematic Singularity of Planar Linkages

Baokun Li and Guangbo Hao

Abstract

The theory of nonlinear stiffness characteristic by employing the kinematic limb-singularity of planar mechanisms with attached springs is proposed. After constructing the position formula with closed-loop form of the mechanism, the kinematic limb-singularity can be identified. The kinetostatic model can be obtained based on the principle of virtual work. The influences of spring stiffness on the force-displacement or torque-angle curve are analysed. Different spring stiffness results in one of four types of stiffness characteristic, which can be used to design an expected stiffness characteristic. After replacing corresponding joints with flexures, the pseudo-rigid-body model of the linkage with springs is obtained. The compliant mechanisms with nonlinear stiffness characteristic can further be synthesised based on the pseudo-rigid-body model.

Keywords: kinematic singularity, nonlinear stiffness, kinetostatic model, planar linkage with springs, compliant mechanism

1. Introduction

A planar linkage always arrives at some several special positions, which may decrease the stability, disable the motion ability or change the degree of freedom of the linkage. These special positions are called kinematic singular configuration or kinematic singularity. It is one of intrinsic properties of the linkage [1].

Kinematic singularity attracted many scholars' attention since it affects the performance of the linkage. Kinematic singularity classification, singularity identification and singularity property, with a particular emphasis on eliminating or avoiding the singularities, are discussed [2–8].

However, everything has two sides. Kinematic singularity of linkages can also be applied to create new useful devices, such as fixture based on the dead-point singularity. In recent years, some compliant mechanisms with new performance are constructed using the kinematic singularity [9–12].

For a generic planar linkage, kinematic limb-singularity and kinematic actuation-singularity may often be exhibited. In this chapter, we mainly introduce how to use the kinematic limb-singularity of the linkage with placing springs at corresponding joints to generate the kinetostatic nonlinear stiffness, which can also be used to synthesise the nonlinear stiffness compliant mechanisms using the pseudo-rigid-body model (PRBM) [13, 14].

2. Position analysis and kinematic singularity identification

Position analysis is the base of the kinematic singularity determination and singularity classification. Many approaches can be used to carry out the position analysis. Graphical methods often provide a fast and efficient means of analysing mechanisms. Analytical methods are currently more common because of the ease in which they are programmed. Here we mainly use the closed-form solution to present the position analysis of the planar linkage followed by the kinematic singularity identification.

By a generic planar double-slider linkage as an example, the position analysis using the closed-form equation and the kinematic singularity identification are introduced.

Consider a planar double-slider linkage with given structure parameters as shown in **Figure 1**.

The right-hand rule fixed frame *O-XYZ* is attached on the base, where the intersection point of two paths of points *A* and *B* is set as the origin *O*. We suppose that the moving direction of the input slider is the negative of *X*-axis. The rotation angle from the negative moving direction of point *A* to the initial moving direction of point *B* is defined by α. Line *AC* is the vertical line from point *A* to line *OB*, where point *C* is the foot. The position vectors of points *A* and *B* with respect to the fixed frame *O-XYZ* are defined by $\boldsymbol{r}_{\mathrm{A}}$ and $\boldsymbol{r}_{\mathrm{B}}$, respectively. The position vector from point *A* to point *B* is defined by $\boldsymbol{r}_{\mathrm{AB}}$. Thus the closed-loop vector equation of the linkage as shown in **Figure 1** can be obtained as

$$\boldsymbol{r}_{\mathrm{A}} + \boldsymbol{r}_{\mathrm{AB}} = \boldsymbol{r}_{\mathrm{B}} \tag{1}$$

The *X*-axis coordinate of point *A* is defined by r_{A}, the rotation angle from vector $\boldsymbol{r}_{\mathrm{A}}$ to vector $\boldsymbol{r}_{\mathrm{B}}$ is defined by θ_{A}, the distance between point *B* and origin *O* is defined by scalar r_{B}, and the length of the coupler *AB* is defined by scalar r_{AB}. Thus, based on Eq. (1), two algebraic equations can be transformed as follows:

$$\begin{cases} r_{\mathrm{A}} + r_{\mathrm{AB}}\cos\theta_{\mathrm{A}} = r_{\mathrm{B}}\cos\alpha \\ r_{\mathrm{AB}}\sin\theta_{\mathrm{A}} = r_{\mathrm{B}}\sin\alpha \end{cases} \tag{2}$$

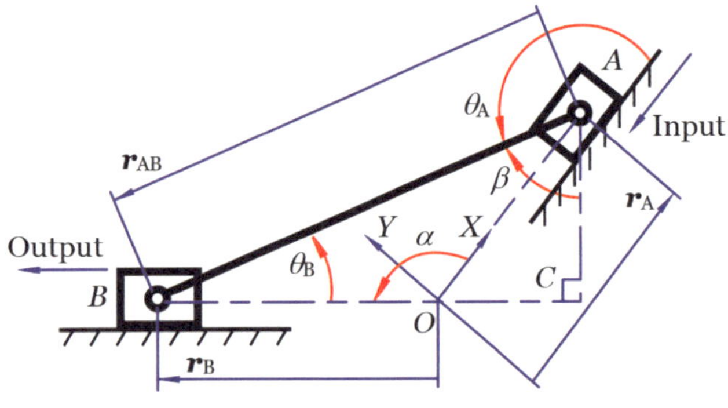

Figure 1.
Double-slider linkage.

Considering the symmetry, the case of $\alpha > 180°$ can be treated as the case of $\alpha < 180°$. Therefore, α is set to satisfy the following condition:

$$0 < \alpha < 180^{\circ} \tag{3}$$

The solution of Eq. (2) for r_B with eliminating θ_A yields

$$r_B = \sqrt{r_{AB}^2 - r_A^2 \sin^2\alpha} + r_A \cos\alpha \tag{4}$$

$$\text{or } r_B = -\sqrt{r_A^2 - r_A^2 \sin^2\alpha} + r_A \cos\alpha \tag{5}$$

where Eq. (5) exists when the output slider is located at the right side of point *C* and moves right. With considering the symmetry, here we only discuss the case that the output slider is located at the left side of point *C* and moves left, which is described by Eq. (4).

The initial distance between origin *O* and point *B* corresponding to Eq. (4) is

$$r_{B0} = \sqrt{r_{AB}^2 - r_{A0}^2 \sin^2\alpha} + r_{A0} \cos\alpha \tag{6}$$

According to Eq. (4), we can further obtain

$$\frac{dr_B}{dr_A} = -\frac{r_A \sin^2\alpha}{\sqrt{r_{AB}^2 - r_A^2 \sin^2\alpha}} + \cos\alpha \tag{7}$$

Eq. (7) shows that if

$$r_A = r_{AB} / \tan\alpha \tag{8}$$

then

$$\frac{dr_B}{dr_A} = 0 \tag{9}$$

It indicates that if $AB \perp OA$, then the linkage is in kinematic limb-singularity which occurs when the instant velocity ratio between the output and the input is equal to zero.

If $r_A = \pm r_{AB}/\sin\alpha$, which occurs when $AB \perp OB$, then

$$\frac{dr_B}{dr_A} = \infty \tag{10}$$

It shows that the instant velocity ratio between the output and the input is equal to infinity, which is the kinematic actuation-singularity. Here we are limited to use the kinematic limb-singularity to design a mechanism with nonlinear stiffness by placing appropriate springs at corresponding joints. Therefore, the coordinate of the input slider should satisfy the kinematic constraint as follows:

$$-r_{AB} / \sin\alpha < r_A < r_{AB} / \sin\alpha \tag{11}$$

In order to pass the kinematic limb-singularity position, the initial coordinate/position of the input slider, r_{A0}, should satisfy

$$r_{A0} > r_{AB} / \tan\alpha \tag{12}$$

3. Force equilibrium equation

We suppose that each of the two prismatic joints is attached a translational spring and each of the two rotational joints is attached a torsional spring. The planar double-slider linkage with springs is obtained as shown in **Figure 2**. The stiffnesses of the translational springs placed at prismatic joints A and B are defined by k_{PA} and k_{PB}, respectively, and the stiffnesses of the translational springs placed at prismatic joints A and B are defined by k_{PA} and k_{PB}, respectively.

The potential energy of the whole mechanism as shown in **Figure 2** can be derived as

$$U = \frac{1}{2}k_{\mathrm{PA}}(r_{\mathrm{A}} - r_{\mathrm{A0}})^2 + \frac{1}{2}(k_{\mathrm{RA}} + k_{\mathrm{RB}})(\theta_{\mathrm{A}} - \theta_{\mathrm{A0}})^2 + \frac{1}{2}k_{\mathrm{PB}}(r_{\mathrm{B}} - r_{\mathrm{B0}})^2 \tag{13}$$

where θ_{A0} is the initial angle between positive direction of X-axis and coupler AB. After differentiating θ_{A}, θ_{A0}, r_{B} and r_{B0}, which can be derived from the geometry of the linkage as shown in **Figure 1**, with respect to r_{A}, and substituting them into Eq. (13), the following can be further derived:

$$\begin{aligned} U = {} & \frac{1}{2}k_{\mathrm{PA}}(r_{\mathrm{A}} - r_{\mathrm{A0}})^2 + \frac{1}{2}(k_{\mathrm{RA}} + k_{\mathrm{RB}})\left(\arccos\frac{r_{\mathrm{A}}\sin\alpha}{r_{\mathrm{AB}}} - \arccos\frac{r_{\mathrm{A0}}\sin\alpha}{r_{\mathrm{AB}}}\right)^2 \\ & + \frac{1}{2}k_{\mathrm{PB}}\left(\sqrt{r_{\mathrm{AB}}^2 - r_{\mathrm{A}}^2\sin^2\alpha} + r_{\mathrm{A}}\cos\alpha - \sqrt{r_{\mathrm{AB}}^2 - r_{\mathrm{A0}}^2\sin^2\alpha} - r_{\mathrm{A0}}\cos\alpha\right)^2. \end{aligned} \tag{14}$$

The principle of virtual work, which does not need to determine the inner forces between two connected links, is a simplified useful method to construct the force equilibrium equation. According to the principle of virtual work, the required driving force, F_{d}, applied on the input slider can be obtained as

$$\begin{aligned} F_{\mathrm{d}} = \frac{dU}{dr_A} = {} & -k_{\mathrm{PA}}(r_{\mathrm{A}} - r_{\mathrm{A0}}) + (k_{\mathrm{RA}} + k_{\mathrm{RB}})\left(\arccos\frac{r_{\mathrm{A}}\sin\alpha}{r_{\mathrm{AB}}} - \arccos\frac{r_{\mathrm{A0}}\sin\alpha}{r_{\mathrm{AB}}}\right)\frac{\sin\alpha}{\sqrt{r_{\mathrm{AB}}^2 - r_{\mathrm{A}}^2\sin^2\alpha}} \\ & + k_{\mathrm{PB}}\left(\sqrt{r_{\mathrm{AB}}^2 - r_{\mathrm{A}}^2\sin^2\alpha} + r_{\mathrm{A}}\cos\alpha - \sqrt{r_{\mathrm{AB}}^2 - r_{\mathrm{A0}}^2\sin^2\alpha} - r_{\mathrm{A0}}\cos\alpha\right)\left(\frac{r_{\mathrm{A}}\sin^2\alpha}{\sqrt{r_{\mathrm{AB}}^2 - r_{\mathrm{A}}^2\sin^2\alpha}} - \cos\alpha\right). \end{aligned} \tag{15}$$

Here the input slider displacement is denoted by S, which satisfies

$$S = -r_{\mathrm{A}} + r_{\mathrm{A0}} \geq 0 \tag{16}$$

where $S \geq 0$ means the input slider moves along the negative direction of X-axis.

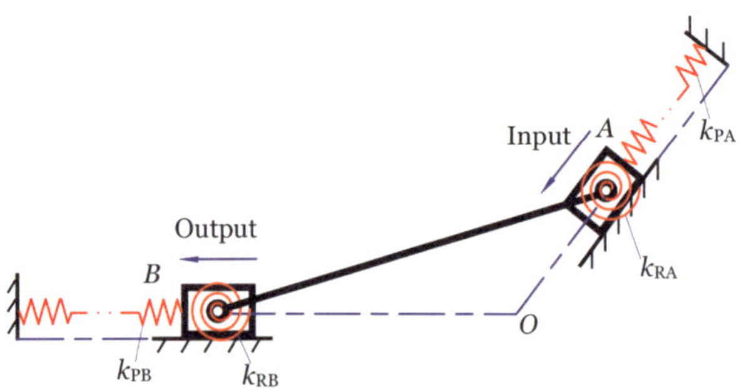

Figure 2.
Planar double-slider linkage with springs.

Substituting Eq. (16) into Eqs. (14) and (15), the potential energy, U, and the driving force, F_d, with respect to S can be obtained as

$$U = \frac{1}{2}k_{\mathrm{PA}}S^2 + \frac{1}{2}(k_{\mathrm{RA}} + k_{\mathrm{RB}})\left(\arccos\frac{(r_{\mathrm{A0}} - S)\sin\alpha}{r_{\mathrm{AB}}} - \arccos\frac{r_{\mathrm{A0}}\sin\alpha}{r_{\mathrm{AB}}}\right)^2$$
$$+\frac{1}{2}k_{\mathrm{PB}}\left(\sqrt{r_{\mathrm{AB}}^2 - (r_{\mathrm{A0}} - S)^2\sin^2\alpha} + (r_{\mathrm{A0}} - S)\cos\alpha - \sqrt{r_{\mathrm{AB}}^2 - r_{\mathrm{A0}}^2\sin^2\alpha} - r_{\mathrm{A0}}\cos\alpha\right)^2. \tag{17}$$

$$F_\mathrm{d} = \frac{dU}{dS} = k_{\mathrm{PA}}S + (k_{\mathrm{RA}} + k_{\mathrm{RB}})\left(\arccos\frac{(r_{\mathrm{A0}} - S)\sin\alpha}{r_{\mathrm{AB}}} - \arccos\frac{r_{\mathrm{A0}}\sin\alpha}{r_{\mathrm{AB}}}\right)\frac{\sin\alpha}{\sqrt{r_{\mathrm{AB}}^2 - (r_{\mathrm{A0}} - S)^2\sin^2\alpha}}$$
$$+ k_{\mathrm{PB}}\left((r_{\mathrm{A0}} - S)\cos\alpha + \sqrt{r_{\mathrm{AB}}^2 - (r_{\mathrm{A0}} - S)^2\sin^2\alpha} - r_{\mathrm{A0}}\cos\alpha - \sqrt{r_{\mathrm{AB}}^2 - r_{\mathrm{A0}}^2\sin^2\alpha}\right)$$
$$\times\left(\frac{(r_{\mathrm{A0}} - S)\ \sin^2\alpha}{\sqrt{r_{\mathrm{AB}}^2 - (r_{\mathrm{A0}} - S)^2\sin^2\alpha}} - \cos\alpha\right). \tag{18}$$

From Eq. (18), we can predict that the variation of driving force exerted on the input slider versus input displacement, i.e., F_d-S curve, would present nonlinear stiffness characteristic.

4. Cause of nonlinear stiffness characteristic generation

It is evident that the springs placed at joints are the cause of nonlinear stiffness characteristic generation, so it is necessary to discuss the influence of spring stiffness on the F_d-S curve characteristic. When the influence of one specific spring stiffness is analysed, every other spring stiffness is set to zero.

4.1 Influences of translational spring stiffness placed at output slider

Substituting $k_{\mathrm{PA}} = 0$ and $k_{\mathrm{RA}} = k_{\mathrm{RB}} = 0$ into Eqs. (17) and (18), the following can be obtained, respectively:

$$U = \frac{1}{2}k_{\mathrm{PB}}\left((r_{\mathrm{A0}} - S)\cos\alpha + \sqrt{r_{\mathrm{AB}}^2 - (r_{\mathrm{A0}} - S)^2\sin^2\alpha} - r_{\mathrm{A0}}\cos\alpha - \sqrt{r_{\mathrm{AB}}^2 - r_{\mathrm{A0}}^2\sin^2\alpha}\right)^2 \tag{19}$$

$$F_\mathrm{d} = \frac{dU}{dS} = k_{\mathrm{PB}}\left((r_{\mathrm{A0}} - S)\cos\alpha + \sqrt{r_{\mathrm{AB}}^2 - (r_{\mathrm{A0}} - S)^2\sin^2\alpha} - r_{\mathrm{A0}}\cos\alpha - \sqrt{r_{\mathrm{AB}}^2 - r_{\mathrm{A0}}^2\sin^2\alpha}\right)$$
$$\times\left(\frac{(r_{\mathrm{A0}} - S)\sin^2\alpha}{\sqrt{L^2 - (r_{\mathrm{A0}} - S)^2\sin^2\alpha}} - \cos\alpha\right). \tag{20}$$

When Eq. (20) is zero, solving this equation with respect to S obtains

$$S_1 = 0, \quad S_2 = -r_{\mathrm{AB}}/\tan\alpha + r_{\mathrm{A0}}, \quad S_3 = -2r_{\mathrm{A0}}\cos^2\alpha - 2\sqrt{r_{\mathrm{AB}}^2 - r_{\mathrm{A0}}^2\sin^2\alpha}\cos\alpha + 2r_{\mathrm{A0}} \tag{21}$$

where S_2 is the kinematic limb-singularity position (based on Eqs. (8) and (16)). From Eq. (19), we know that

$$U|_{S=S_1} = U|_{S=S_3} = 0 \tag{22}$$

Substitution of Eq. (16) into the differentiation of Eq. (20) with respect to S leads to

$$\left.\frac{dF_{\mathrm{d}}}{dS}\right|_{S=S_2} = \left.\frac{d^2U}{dS^2}\right|_{S=S_2} = -k_{\mathrm{PB}}\left(\frac{r_{\mathrm{AB}}}{\sin\alpha} - r_{\mathrm{B0}}\right)\cdot\frac{1}{r_{\mathrm{AB}}\sin\alpha} < 0. \tag{23}$$

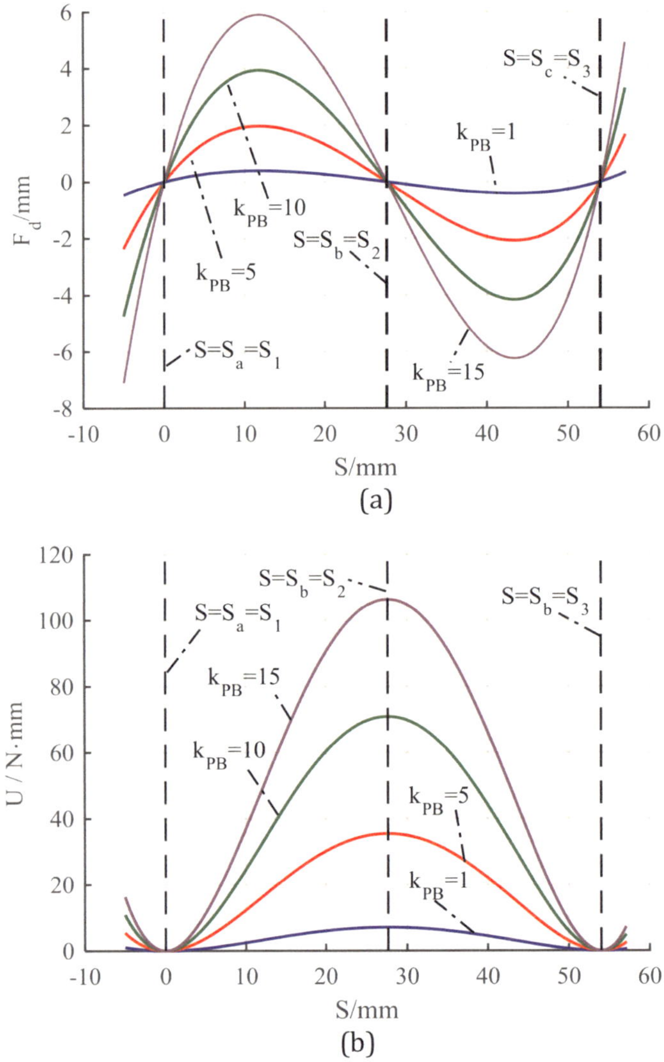

Figure 3.
The bistable characteristic with different k_{PB} when $k_{PA} = k_{RA} = 0$ and $k_{RB} = 0$. (a) Driving force versus input displacement. (b) Potential energy versus input displacement.

Eq. (23) shows that when $S = S_2$, the potential energy, U, reaches the local maximum. Therefore, $S = S_2$, the kinematic limb-singularity position, is also the unstable equilibrium position [15].

Therefore, according to [15], when $k_{PA} = 0$, $k_{RA} = k_{RB} = 0$ and $k_{PB} \neq 0$, the kinematic limb-singularity position, S_2, is the mechanism's unstable equilibrium position and S_1 and S_3 are the stable equilibrium points.

If the coupler length, r_{AB}, is 100 mm; intersection angle, α, is 100°; initial input slider position, r_{A0}, is 10 mm; and the unit of k_{PB} is N/mm (here the unit of translational spring stiffness is N/mm, and the unit of torsional spring stiffness is N·mm/rad); the stiffness characteristic produced by the mechanism is shown in **Figure 3**.

Figure 3 confirms that when the spring stiffness placed at output slider is exclusively zero, the mechanism produces the bistable characteristic.

4.2 Influences of translational spring stiffness placed at input slider

Substitution of $k_{PB} = 0$ and $k_{RA} = k_{RB} = 0$ into Eqs. (17) and (18) obtains the expressions as follows:

$$U = \frac{1}{2} k_{PA} S^2 \tag{24}$$

$$F_d = k_{PA}\, S \tag{25}$$

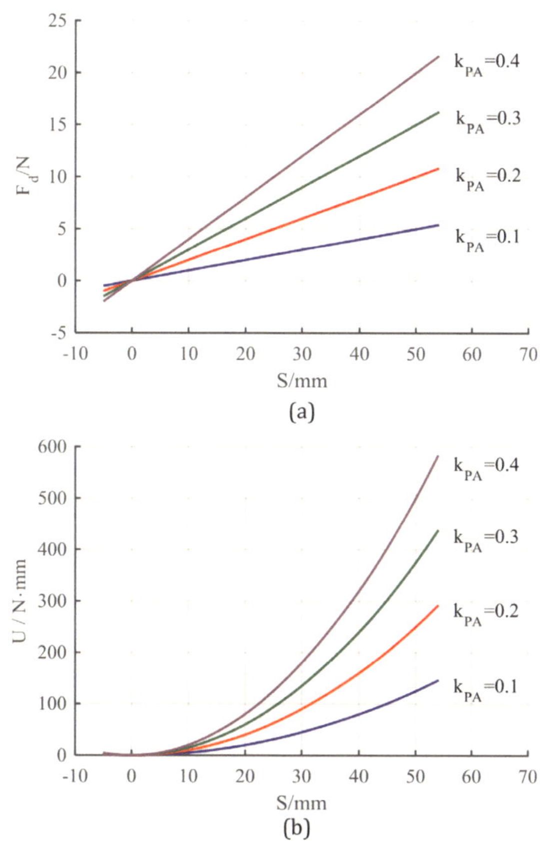

Figure 4.
Stiffness characteristic with different k_{PA} when $k_{RA} = k_{RB} = k_{PB} = 0$. (a) Driving force versus input displacement. (b) Potential energy versus input displacement.

Eqs. (24) and (25) show that the mechanism only generates positive-stiffness characteristic when the mechanism has only one minimal potential energy point, which can be confirmed by **Figure 4**, where specific parameters are given as r_{AB} = 100 mm, α = 100° and r_{A0} = 10 mm.

4.3 Influences of torsional spring stiffness placed at pin joints

When $k_{PA} = k_{PB} = 0$ and $k_{RA} = k_{RB} \neq 0$, from Eq. (18), we can obtain

$$F_d = (k_{RA} + k_{RB})\left(\arccos\frac{(r_{A0} - S)\sin\alpha}{r_{AB}} - \arccos\frac{r_{A0}\sin\alpha}{r_{AB}}\right)\frac{\sin\alpha}{\sqrt{r_{AB}^2 - (r_{A0} - S)^2\sin^2\alpha}} \quad (26)$$

where only and only if $S = S_1 = 0$, i.e., $r_A = r_{A0}$, then $F_d = 0$.

In other words, if and only if $S = S_1 = 0$, the mechanism is in equilibrium position without external force. When the mechanism is located in any other positions, it is unstable except when applied by external force. Meanwhile, the potential energy, U, has no local maximum but has only one minimum which is located at $S = S_1 = 0$.

For $k_{PA} = k_{PB} = 0$, $k_{RA} = k_{RB} \neq 0$, when r_{AB} = 100 mm, α = 100° and r_{A0} = 10 mm, the force-displacement characteristic and the potential energy curve are shown in **Figure 5**.

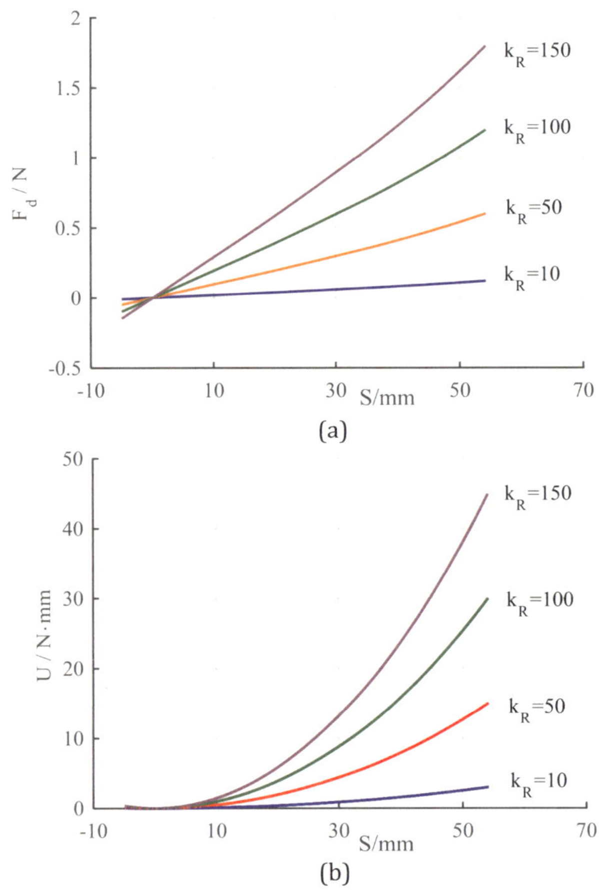

Figure 5.
Behaviours with different $k_{RA} = k_{RB} = k_R$ *when* $k_{PB} = 0$ *and* $k_{PA} = 0$. *(a) Driving force versus input displacement. (b) Potential energy versus input displacement.*

Figure 5 demonstrates that when the torsional springs are placed at the pin joints, the mechanism only produces positive-stiffness characteristic but does not produce other stiffness characteristic.

5. Nonlinear stiffness characteristic construction

Section 4 showed that spring placed at the output slider causes the bistable characteristic with the negative domain, while springs placed at other joints produce the corresponding positive-stiffness characteristic. It can be predicted that if more than one spring are placed at joints, when the mechanism moves from non-singular position (**Figure 6(a)**) to another non-singular position (**Figure 6(c)**) while passing through the limb-singularity position (**Figure 6(b)**), the stiffness characteristic of the mechanism is the superposition of the corresponding stiffness characteristic caused by the joints. For instance, if $k_{PB} = 1$ N/mm, $\alpha = 100°$,

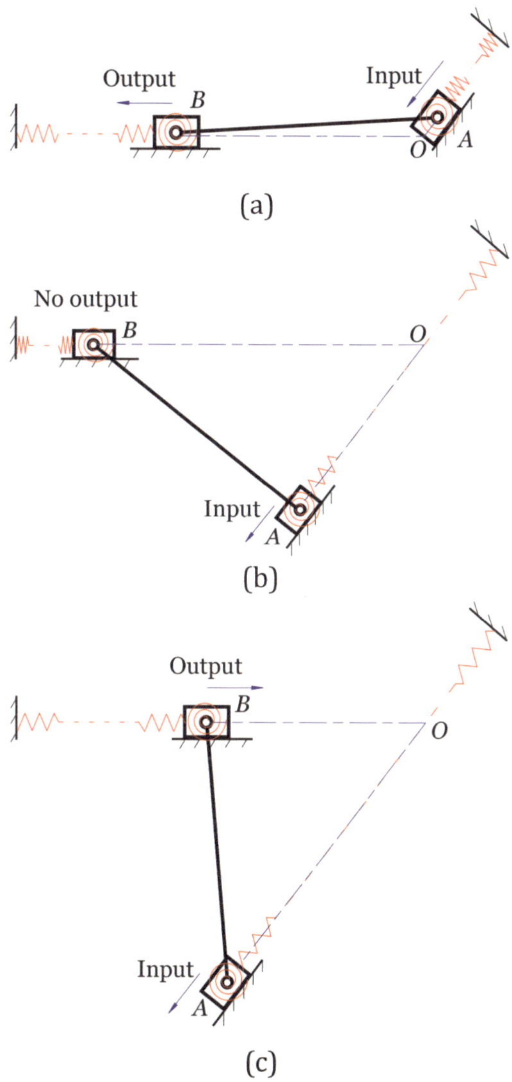

Figure 6.
Positions of the mechanism. (a) Initial kinematic non-singular position. (b) Kinematic limb-singularity position. (c) End kinematic non-singular position.

r_{A0} = 10 mm and r_{AB} = 100 mm, several nonlinear stiffness characteristics with different spring stiffness are shown in **Figure 7**.

Figure 7 shows that after assigning appropriate spring stiffness placed at the corresponding joints, the mechanism can generate one of four types of nonlinear

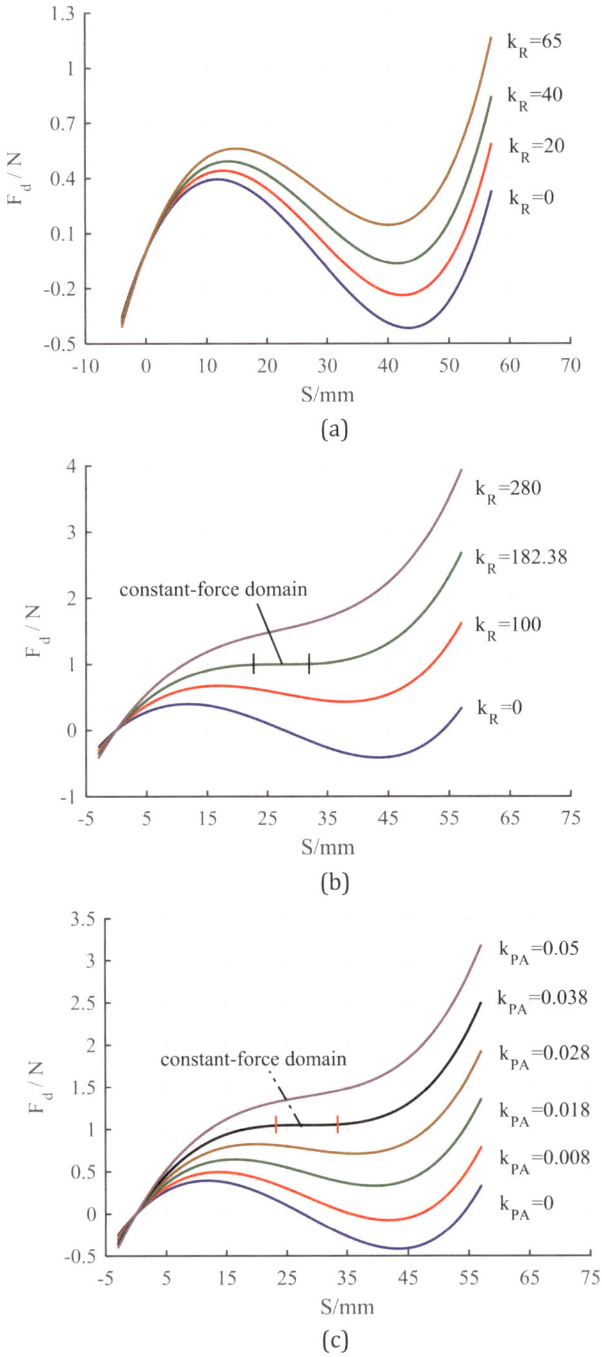

Figure 7.
Nonlinear stiffness characteristic when k_{PB} = 1 N/mm. (a) Driving force versus input displacement for minor change of k_{RA} = k_{RB} = k_R when k_{PA} = 0. (b) Driving force versus input displacement for large change of k_{RA} = k_{RB} = k_R when k_{PA} = 0. (c) Driving force versus input displacement for different spring stiffness when k_{RA} = k_{RB} = 0.

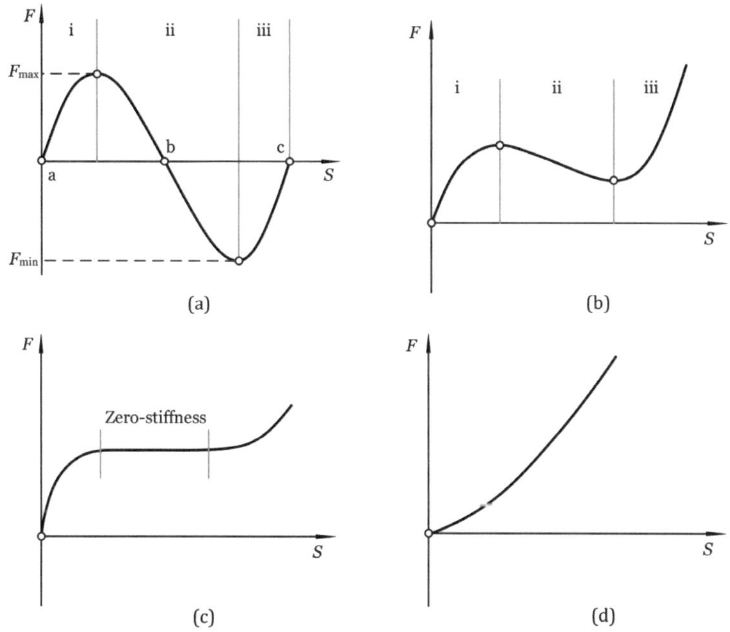

Figure 8.
Four types of nonlinear stiffness characteristic of the mechanism. (a) Bistable characteristic, (b) partial negative-stiffness characteristic, (c) partial zero-stiffness characteristic, and (d) positive-stiffness characteristic.

stiffness characteristics including bistable characteristic, partial negative-stiffness characteristic, partial zero-stiffness characteristic and positive-stiffness characteristic which are shown in **Figure 8**, as it works around the kinematic limb-singularity position.

The above-mentioned analysis illustrates the case that the mechanism moves from a non-singular position. When the mechanism moves from the kinematic limb-singularity position (**Figure 6(b)**) to a non-singular position (**Figure 6(c)**), every spring transforms with zero potential energy to a position with a certain amount of potential energy. The total potential energy of the mechanism starts from zero to nonzero without local minimal energy point except the initial position. Every spring force/torque increases in the process of the mechanism's motion. From Eqs. (18) or (20), the driving force is to overcome the resistance caused by every spring, so the driving force increases when the mechanism moves from non-singular position. In other words, when mechanism moves from non-singular position with no deflected springs, the mechanism only generates the positive-stiffness characteristic as shown in **Figure 8(d)**. If $k_{PA} = k_{PB} = 0$, $k_{RA} = k_{RB} \neq 0$ and $r_{AB} = 100$ mm, $\alpha = 100°$ and $r_{A0} = 10$ mm, the stiffness characteristic is shown in **Figure 9**.

Figure 9 demonstrates that when the mechanism starts from the kinematic limb-singularity position with no deflected springs towards a non-singular position, it only generates positive-stiffness characteristic.

As the final stiffness characteristic is determined by the superposition of stiffness characteristic caused by each spring, an expected stiffness characteristic can be constructed by assigning appropriate values to k_{PB}, k_{RA}, k_{RB} and k_{PA} on the condition of $k_{RB} \neq 0$ when the mechanism moves from one non-singular position to another non-singular position with passing through the kinematic limb-singularity position. The method for designing an expected nonlinear stiffness design is proposed in [16].

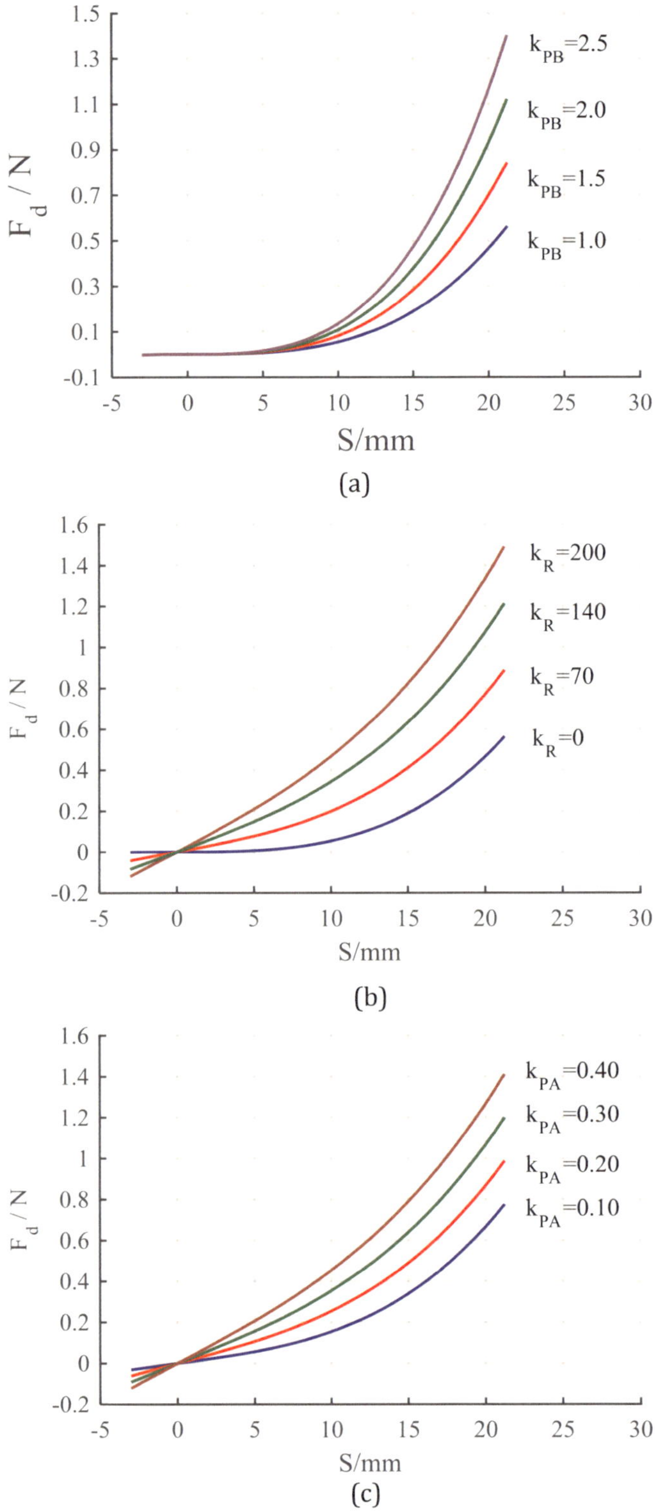

Figure 9.
Stiffness characteristic with initial non-singular position. (a) Stiffness characteristic for different k_{PB} *when* $k_{RA} = k_{RB} = 0$ *and* $k_{PA} = 0$. *(b) Stiffness characteristic for different* $k_{RA} = k_{RB} = k_R$ *when* $k_{PB} = 1$ *N/mm and* $k_{PA} = 0$. *(c) Stiffness characteristic for* k_{PA} *when* $k_{PB} = 1$ *N/mm and* $k_{RA} = k_{RB} = 0$.

6. Further discussion

From Sections 2–5, it can be shown that if a planar linkage exhibits a kinematic limb-singularity, it generates different nonlinear spring stiffness characteristics if the linkage is added with springs at corresponding joints with the condition that the spring stiffness corresponding to output slider is nonzero. Nonlinear stiffness characteristic generation using the kinematic limb-singularity of a planar linkage can be demonstrated by another planar linkage with springs as shown in **Figure 10**, which represents a crank-slider linkage with springs.

The Cartesian coordinates system, *O-xyz*, is constructed as shown in **Figure 10**, where crank *AB* rotates about joint *A* in an anticlockwise direction, the slider moves along the *x*-axis, and coupler *BC* connects link *AB* and slider by two rotation joints *B* and *C*. The three rotation joints are added with torsional springs with spring stiffnesses denoted by K_{RA}, K_{RB} and K_{RC}, respectively. An extension spring is placed at the output slider and its stiffness is denoted by K_{PC}.

The position formula with closed-loop form, whose derivation process is similar to one of the double-slider four-bar linkages, can be established easily (not shown here).

Based on the position analysis, the kinetostatic model of the mechanism by using the principle of virtual work can be constructed as

$$\begin{aligned} T_{\mathrm{d}} = {} & K_{\mathrm{RA}}(\theta_{\mathrm{A}} - \theta_{\mathrm{A0}}) + K_{\mathrm{RB}}\left(-\theta_{\mathrm{A}} - \arcsin\frac{r_1 \sin\theta_{\mathrm{A}} - e}{r_2} + \theta_{\mathrm{A0}} + \arcsin\frac{r_1 \sin\theta_{\mathrm{A0}} - e}{r_2}\right) \\ & \times(-1 - r_1\cos\theta_{\mathrm{A}}/a) \\ & + K_{\mathrm{RC}}\left(\arcsin\frac{r_1 \sin\theta_{\mathrm{A}} - e}{r_2} - \arcsin\frac{r_1 \sin\theta_{\mathrm{A0}} - e}{r_2}\right) \times \frac{r_1\cos\theta_{\mathrm{A}}}{a} \\ & + K_{\mathrm{PC}}(r_1\cos\theta_{\mathrm{A}} + a - r_1\cos\theta_{\mathrm{A0}} - a_0) \times (-r_1\sin\theta_{\mathrm{A}} - b/a) \end{aligned} \tag{27}$$

where r_1 and r_2 are crank length and coupler length, respectively, r_3 is the *X*-axis coordinate of output slider and *e* is the offset.

Moreover, *a* and a_0 are defined by

$$a = \sqrt{r_2^2 - (r_1\sin\theta_{\mathrm{A}} - e)^2}, a_0 = \sqrt{r_2^2 - (r_1\sin\theta_{\mathrm{A0}} - e)^2}$$

According to Eq. (27), the T_d-θ_A (driving torque versus input position angle) curve can be described.

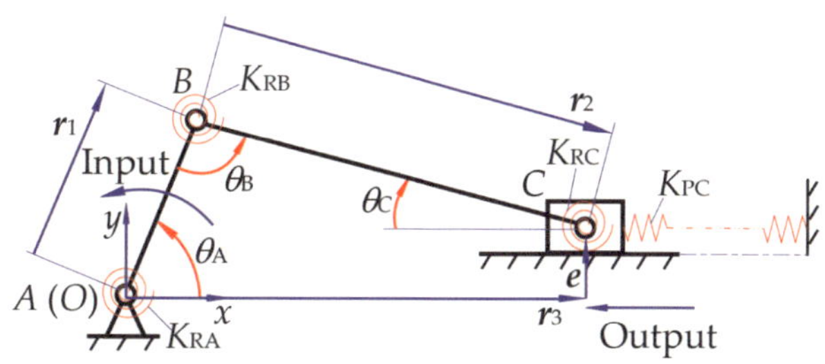

Figure 10.
Crank-slider with springs.

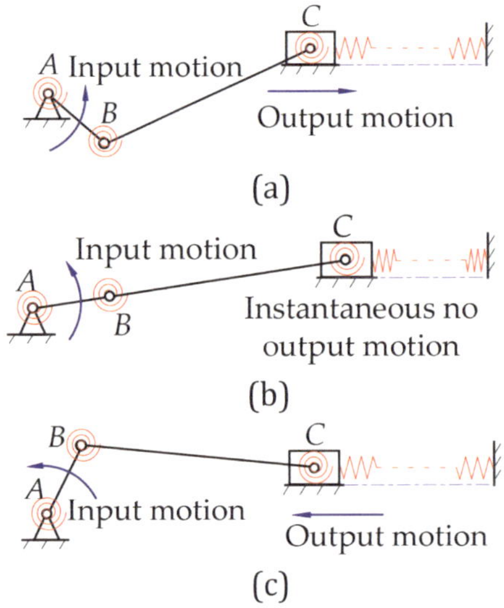

Figure 11.
Different positions of the crank-slider mechanism with springs. (a) Non-singular initial position. (b) Kinematic limb-singularity position. (c) Non-singular end position.

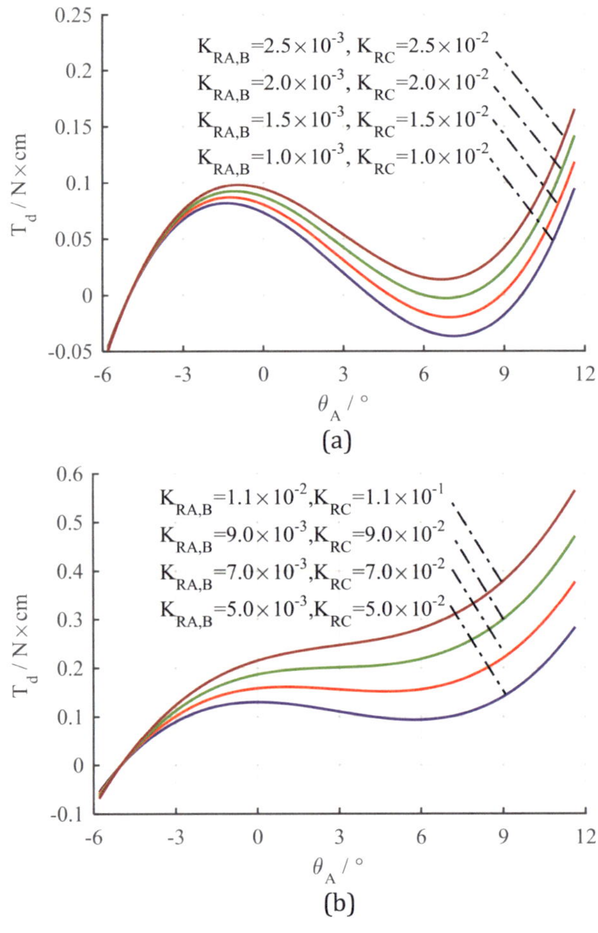

Figure 12.
Case that the mechanism moves from non-singular position. (a) Input torque variation when $K_{RA} = K_{RB} = K_R$ *is small. (b) Input torque variation when* $K_{RA} = K_{RB} = K_R$ *is large.*

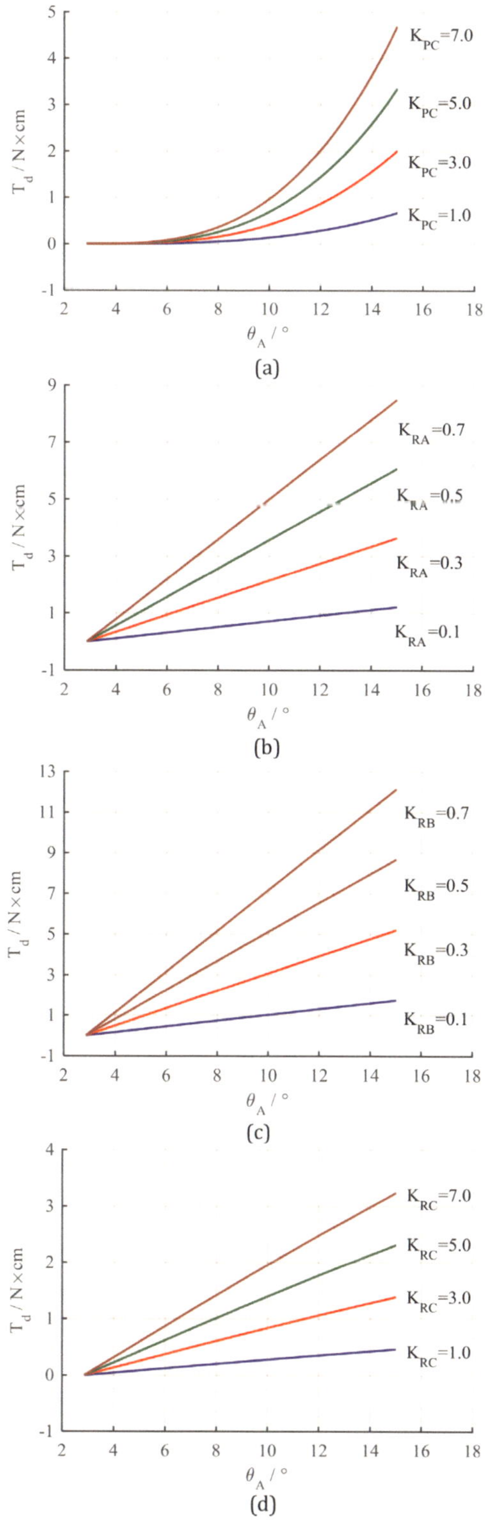

Figure 13.
Case that the mechanism moves from kinematic limb-singularity position. (a) Case of different K_{PC} *when* $K_{RA} = K_{RB} = K_{RC} = 0$*. (b) Case of different* K_{RA} *when* $K_{RB} = K_{RC} = 0$ *and* $K_{PC} = 0$*. (c) Case of different* K_{RB} *when* $K_{RA} = K_{RC} = 0$ *and* $K_{PC} = 0$*. (d) Case of different* K_{RC} *when* $K_{RA} = K_{RB} = 0$ *and* $K_{PC} = 0$*.*

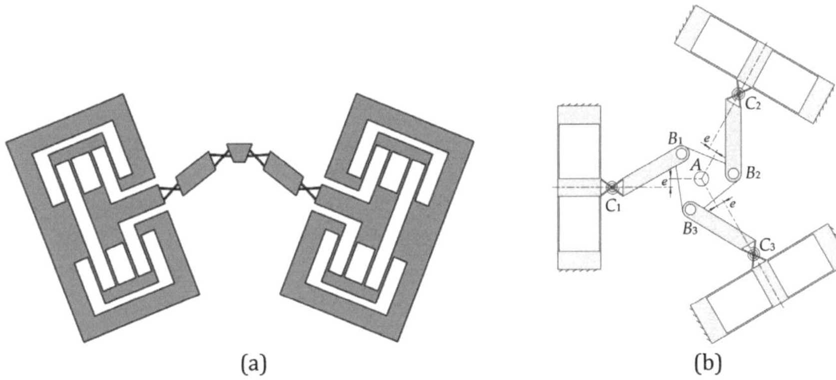

Figure 14.
Nonlinear stiffness characteristic compliant mechanisms based on the PRBM. (a) Compliant double-slider mechanism and (b) compliant crank-slider mechanism.

When the mechanism with springs moves from one non-singular position to the kinematic limb-singularity position and then stops at another non-singular position as shown in **Figure 11**, it may produce one of four types of nonlinear stiffness characteristics including the bistable characteristic, partial negative-stiffness characteristic, partial zero-stiffness characteristic and positive-stiffness characteristic. For illustration, in **Figure 12**, crank length, r_1, is 10 cm; coupler length, r_2, is 50 cm; offset e is 3 cm; input initial position angle, θ_{A0}, is $-5°$; and K_{PC} = 1 N/cm.

It can also be shown that, similarly to the double-slider four-bar mechanism with springs, when the crank-slider mechanism with springs moves from the kinematic limb-singularity position, it only generates the positive-stiffness characteristic, which is shown in **Figure 13**, where the geometry parameters are given as the same as shown in **Figure 12**.

Therefore, we can conclude that after placing springs at different pair combinations, a planar linkage which has the kinematic limb-singularity can generate corresponding nonlinear stiffness characteristic in condition that the mechanism moves from initial non-singular position with no deflected springs to the kinematic limb-singularity position and then stops at another non-singular position. If the mechanism moves from the kinematic limb-singularity position with no deflected springs, it only generates the positive-stiffness characteristic.

It is worth to point out that the nonlinear stiffness characteristic generation method can also be applied to design the nonlinear stiffness characteristic compliant mechanism by using the PRBM as shown in **Figure 14**.

In **Figure 14**, the equivalent stiffness of compliant rotational joint and compliant translational joint can be calculated by referring to previous work [17, 18].

7. Conclusions

The kinematic limb-singularity positions of planar linkages with attached springs can be used to generate nonlinear characteristics. After assigning different spring stiffness, the mechanism may exhibit one of four types of nonlinear stiffness characteristics. These are the bistable characteristic, partial negative-stiffness characteristic, partial zero-stiffness characteristic and positive-stiffness characteristic. The type of stiffness characteristic is determined by the motion model and the value of spring stiffness. On the condition that the mechanism moves from the initial non-singular position to another non-singular position with passing though the kinematic limb-singularity position, spring stiffness determines one of four types of stiffness characteristics. On the other hand, when the mechanism moves from the kinematic limb-singularity position, it only produces the positive-stiffness characteristic.

Acknowledgements

The authors gratefully acknowledge the support of the National Natural Science Foundation of China under Grant No. 51605006, the Research Foundation of Key Laboratory of Manufacturing Systems and Advanced Technology of Guangxi Province, China, under Grant No. 17-259-05-013K.

Author details

Baokun Li[1] and Guangbo Hao[2*]

1 School of Mechanical Engineering, Anhui University of Science and Technology, Huainan, China

2 School of Engineering, University College Cork, Cork, Ireland

*Address all correspondence to: g.hao@ucc.ie

References

[1] Huang Z, Zhao YS, Zhao TS. Advanced Spatial Mechanism. Beijing, China: Higher Education Press; 2005. (in Chinese)

[2] Gosselin CM, Angeles J. Singularity analysis of closed loop kinematic chains. IEEE Transactions on Robotics and Automation. 1990;**6**(3):281-290. DOI: 10.1109/70.56660

[3] Amine S, Mokhiamar O, Caro S. Classification of 3T1R parallel manipulators based on their wrench graph. ASME Journal of Mechanisms and Robotics. 2017;**9**(1):011003. DOI: 10.1115/1.4035188

[4] Huang Z, Cao Y. Property identification of the singularity loci of a class of Gough-Stewart manipulators. The International Journal of Advanced Robotics Research. 2005;**24**(8):375-685. DOI: 10.1177/0278364905054655

[5] Boudreau R, Nokleby S. Force optimization of kinematically-redundant planar parallel manipulators following a desired trajectory. Mechanism and Machine Theory. 2012; **56**(10):138-155. DOI: 10.1016/j. mechmachtheory.2012.06.001

[6] Saglia J, Dai JS, Caldwell DG. Geometry and kinematic analysis of a redundantly actuated parallel mechanism that eliminates singularity and improves dexterity. ASME Journal of Mechanical Design. 2008;**130**(12): 124501. DOI: 10.1115/1.2988472

[7] Li B, Cao Y, Zhang Qiuju, et al. Position-singularity analysis of a special class of the Stewart parallel mechanisms with two dissimilar semi-symmetrical hexagons. Robotica. 2013;**31**(1):123-136. DOI: 10.1017/S0263574712000148

[8] Karimia A, Masouleha MT, Cardoub P. Avoiding the singularities of 3-RPR parallel mechanisms via dimensional synthesis and self-reconfigurability. Mechanism and Machine Theory. 2016; **99**:189-206. DOI: 10.1016/j. mechmachtheory.2016.01.006

[9] Abhilash N, Li H, Hao G, et al. A reconfigurable compliant four-bar mechanism with multiple operation modes. In: ASME International Design Engineering Technical Conferences & Computers and Information in Engineering Conference; August 6–9, 2017; Cleveland, Ohio, USA; DETC2017–67441; 2017. DOI: 10.1115/ DETC2017-67441

[10] Rubbert L, Caro S, Gangloff J, et al. Using singularities of parallel manipulators for enhancing the rigid-body replacement design method of compliant mechanisms. ASME Journal of Mechanical Design. 2014;**136**(5): 051010. DOI: 10.1115/1.4026949

[11] Rubbert L, Renaud P, Caro S, et al. Design of a compensation mechanism for an active cardiac stabilizer based on an assembly of planar compliant mechanisms. Mechanics and Industry. 2014;**15**(2):147-151. DOI: 10.1051/meca/ 2014013

[12] Quentin B, Marc V, Salih A. Parallel singularities for the design of softening springs using compliant mechanisms. In: ASME International Design Engineering Technical Conferences & Computers and Information in Engineering Conference; August 2–5, 2015; Boston, Massachusetts, USA; DETC2015-47240; 2015. DOI: 10.1115/DETC2015-47240

[13] Gallego JA, Herder J. Synthesis methods in compliant mechanisms: An overview. In: ASME International Design Engineering Technical Conferences and Computers and Information in Engineering Conference & Computers and Information in Engineering Conference; August

30–September 2, 2009; San Diego, California, USA; DETC2009–86845; 2009. DOI: 10.1115/DETC2009-86845

[14] Olsen BM, Issac Y, Howell LL, et al. Utilizing a classification scheme to facilitate rigid-body replacement for compliant mechanism design. In: ASME International Design Engineering Technical Conferences & Computers and Information in Engineering Conference; August 15–18, 2010; Montreal, Quebec, Canada; DETC2010–28473; 2010. DOI: 10.1115/DETC2010-28473

[15] Baker MS, Howell LL. On-chip actuation of an in-plane compliant bistable micromechanism. Journal of Microelectromechanical Systems. 2002; **11**(5):566-573. DOI: 10.1109/JMEMS.2002.803284

[16] Li B, Hao G. Nonlinear behaviour design using the kinematic singularity of a general type of double-slider four-bar linkage. Mechanism and Machine Theory. 2018;**129**:106-130. DOI: 10.1016/j.mechmachtheory.2018.07.016

[17] Zhao H, Bi S, Yu J. Nonlinear deformation behavior of a beam-based flexural pivot with monolithic arrangement. Precision Engineering. 2011;**35**(2):369-382. DOI: 10.1016/j.precisioneng.2010.12.002

[18] Pei X, Yu J, Zong G, et al. The modeling of cartwheel flexural hinges. Mechanism and Machine Theory. 2009; **44**(10):1900-1909. DOI: 10.1016/j.mechmachtheory.2009.04.006

Section 3

Kinematics for Spacecrafts and Satellites

Chapter 5

Kinematic Absolute Positioning with Quad-Constellation GNSS

Lin Pan, Changsheng Cai, Jianjun Zhu and Xianqiang Cui

Abstract

The absolute positioning technique is based on a point positioning mode with a single Global Navigation Satellite System (GNSS) receiver, which has been widely used in many fields such as vehicle navigation and kinematic surveying. For a long period, this positioning technique mainly relies on a single GPS system. With the revitalization of Global Navigation Satellite System (GLONASS) constellation and two newly emerging constellations of BeiDou Navigation Satellite System (BDS) and Galileo, it is now feasible to carry out the absolute positioning with quad-constellation of GPS, GLONASS, BDS, and Galileo. A combination of multi-constellation observations can offer improved reliability, availability, and accuracy for position solutions. In this chapter, combined GPS/GLONASS/BDS/Galileo point positioning models for both traditional single point positioning (SPP) and precise point positioning (PPP) are presented, including their functional and stochastic components. The traditional SPP technique has a positioning accuracy at a meter level, whereas the PPP technique can reach an accuracy of a centimeter level. However, the later relies on the availability of precise ephemeris and needs a long convergence time. Experiments were carried out to assess the kinematic positioning performance in the two different modes. The positioning results are compared among different constellation combinations to demonstrate the advantages of quad-constellation GNSS.

Keywords: kinematic positioning, Global Navigation Satellite System, multi-constellation combination, single point positioning, precise point positioning

1. Introduction

Position services have become an inevitable demand for the human activities. Advanced technologies of the position services can significantly improve human's manufacturing efficiency, life quality, and resource utilization. Along with the development of human society, there is an increasing need of kinematic position services, such as automatic drive, intelligent transportation, precision agriculture, and so on. The Global Navigation Satellite System (GNSS), which rose in the 1980s of the last century, is an optimal infrastructure to realize the outdoor kinematic position services. The GNSS-based absolute positioning technologies have many advantages, such as no restriction by the inter-station distance, low cost, and simple data processing. The kinematic positions can be derived globally in all weather and any time based on the GNSS absolute positioning technique with a single receiver.

According to different performance demands, two kinds of kinematic absolute positioning technologies can be employed, namely, single point positioning (SPP)

and precise point positioning (PPP). The SPP technology can provide meter-level positioning, while the PPP technology has a positioning accuracy at a centimeter level. As satellite-based positioning technologies, the performances of the SPP and PPP are quite dependent on the observed satellites. For a long period, the kinematic absolute positioning technologies are mainly based on a single GPS system. With the recent revitalization of the Global Navigation Satellite System (GLONASS) constellation and two newly emerging constellations of BeiDou Navigation Satellite System (BDS) and Galileo, the quad-constellation integrated absolute positioning has become feasible. Multi-constellation combination is expected to improve the reliability, availability, and accuracy of the SPP and PPP solutions due to the increased measurement redundancy and enhanced satellite geometry, especially when they are performed in areas with GNSS signal blockages.

2. Global Navigation Satellite System

A rapid development has been undergone for the satellite-based global navigation systems in recent years. The GNSS family has expanded from a single GPS constellation to four constellations of Galileo, BDS, GLONASS, and GPS. An overview of the four GNSS systems is conducted in terms of their space segment status and navigation signals.

2.1 Space segment status

A nearly circular orbit with an altitude of about 20,200 km is employed for GPS satellites. The GPS satellites pass a same place twice a day. All GPS satellites are located on six equally spaced orbital planes surrounding the earth. For each plane, there are four slots occupied by baseline satellites. The ascending nodes of the orbital planes are equally spaced 60° apart, and they are inclined at 55°. From virtually any point on the earth, users can view at least four GPS satellites, attributing to the 24-slot arrangement. Currently, a 27-slot constellation with improved coverage in most parts of the world is effectively operated for the GPS after expanding the constellation. The United States is committed to maintaining the availability of at least 24 operational GPS satellites 95% of the time. For the past few years, a total of 31 operational GPS satellites have been flying so as to ensure this commitment. The GPS constellation is a mix of old and new satellites, including 1 Block IIA, 11 Block IIR, 7 Block IIR-M, and 12 Block IIF satellites, as of March 2019 [1].

A complete revitalization of GLONASS with a full constellation including 24 operational satellites arranged into 3 orbital planes has been achieved since 2012. A nearly circular orbit is operated for each GLONASS satellite at an altitude of about 19,100 km, and approximately 11 h 16 min is needed for the GLONASS satellites to complete the orbit. The ascending nodes of orbital planes are separated by 120°, and each orbital plane has an inclination angle of 64.8°. The satellites within the same orbital plane are equally spaced 45° apart, while the difference in argument of latitude for satellites in equivalent slots in two different orbital planes is 15°. A continuous global navigation and position service can be provided due to the reasonable spacing of GLONASS satellites. Currently, there are 26 GLONASS satellites in orbit, but only 23 of them are in full operation, including 1 GLONASS-K1 and 22 GLONASS-M satellites. The GLONASS-M satellite SVN 716 is spare, and the SVN 720 satellite of the same series is in maintenance. The GLONASS-K1 satellite SVN 701 K is in the phase of flight tests [2].

Following the deployment timeline of BDS, its implementation has been performed in three steps: BeiDou navigation demonstration system (BDS-1) by 2000, regional BDS (BDS-2) by 2012, and global BDS (BDS-3) by 2020. Although the BDS-3 is under construction based on the "three-step" strategy, it has been providing basic services of positioning, navigation, and timing (PNT) for global customers since 27 December 2018. The nominal space constellation of BDS-2 consists of five geostationary earth orbit (GEO) satellites, five inclined geosynchronous orbit (IGSO) satellites, and four medium earth orbit (MEO) satellites. The nominal space constellation of BDS-3 consists of 3 GEO satellites, 3 IGSO satellites, and 24 MEO satellites. The five BDS-2 GEO satellites are located at 58.75°E, 80°E, 110.5°E, 140°E, and 160°E respectively, while the three BDS-3 GEO satellites are placed at 80°E, 110.5°E, and 140°E above the earth's equator, respectively. Actually, spare satellites may be deployed in orbit, according to actual situation. An altitude of 35,786 km with a period of revolution of 23 h 56 min is adopted for the operation of the GEO satellites. The GEO satellites exhibit a non-zero inclination of 0.7–1.7°, as they are actively controlled in longitudes rather than latitudes. The altitude and orbital period of the IGSO satellites are the same as those of the GEO satellites, and they have an inclination of 55°. Each MEO satellite operates in a nearly circular orbit at an orbit inclination of 55° and an altitude of 21,528 km. The MEO satellites are arranged into three orbital planes, and an angle of 120° is used for the spacing between ascending nodes of different orbital planes. For MEO satellites, the period of the revolution is 12 h 53 min. As of March 2019, there are 15 BDS-2 satellites (5 GEO/7 IGSO/3 MEO), 5 BDS-3 demonstration system (BDS-3S) satellites (2 IGSO/3 MEO), and 19 BDS-3 satellites (1 GEO/18 MEO) in orbit. All BDS-3S satellites and the BDS-3 GEO satellite C59 are in the flight test phase, while the other BDS satellites are fully operational [3].

When fully deployed, the Galileo constellation will contain 30 satellites in 3 orbital planes. There are one inactive spare satellite and nine equally spaced operational satellites in each plane. The ascending nodes of the three planes are equally separated by 120°, and all of them are inclined at an angle of 56°. With a period of about 14 h 7 min and a semimajor axis of 29,600 km, all Galileo satellites are in a nearly circular orbit. The current Galileo space segment is composed of 26 satellites of 2 different generations. Two respective dual launches of four in-orbit validation (IOV) satellites in 2011 and 2012 initiated the buildup of the operational Galileo constellation. A permanent failure of the E5 and E6 signal transmission in May 2014 happened for the IOV-4 satellite due to a sudden power loss. Since then, the IOV-4 satellite can only transmit E1 signal. On 22 August 2014, the first pair of Full Operational Capability (FOC) satellites was launched. However, there was an "orbital injection anomaly" for the two FOC satellites 1 day later, which results in an elliptical orbit with an inclination roughly 5° smaller than planned. Because of the lack of broadcast ephemerides and single-frequency transmission, the IOV-4 satellite currently has the status "not available." As of June 2016, the other three IOV satellites are declared "available," namely, providing broadcast ephemerides and healthy signals that can be used in real-time navigation. Although the two FOC satellites FOC-1 and FOC-2 are not listed in the constellation status and in the eccentric orbit, they are generally transmitting broadcast ephemerides and navigation signals. The other 20 FOC satellites were successively declared "available" over the past 4 years [4].

Figure 1 shows a 24-h ground track of quad-constellation GNSS satellites on 18 March 2019. The quad-constellation mixed precise satellite orbit file is used to derive the satellite coordinates for all GNSS satellites. Different satellites are identified by different colors. The coordinate transformation from reference frame "IGS08" (three-dimensional positions x, y, and z), namely, the realization of international terrestrial reference frame 2008, to geodetic coordinate system (geodetic

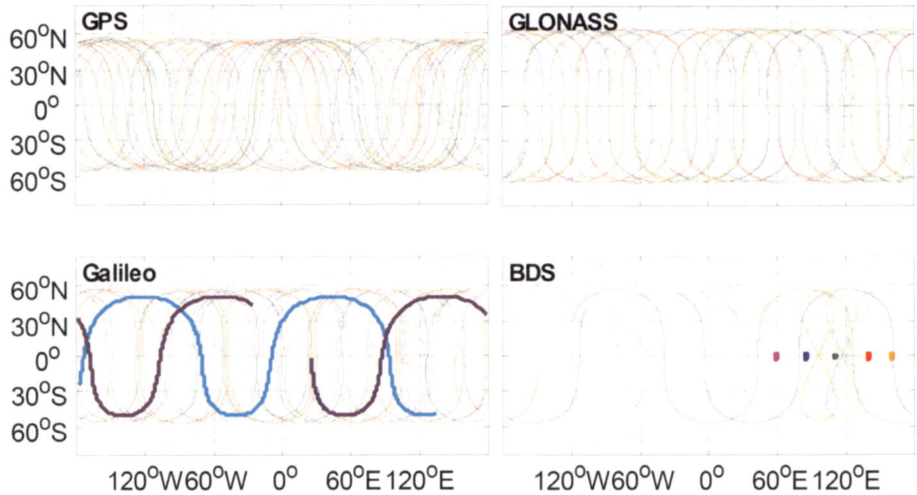

Figure 1.
Ground tracks of quad-constellation navigation systems on 18 March 2019.

latitude *B*, longitude *L*, and height *H*) is first performed. Then, the tracks of GNSS satellites are projected to the grounds in the map based on their geodetic coordinates of latitudes and longitudes. The GNSS satellites can provide better PNT services for the stations near their ground tracks. All MEO satellites offer complete global coverage, including Galileo satellites, BDS MEO satellites, GLONASS satellites, and GPS satellites. The ground tracks of the four types of MEO satellites are confined from 57.2°S to 57.2°N latitude, from 56.1°S to 56.1°N latitude, from 65.6°S to 65.6°N latitude, and from 56.8°S to 56.8°N latitude, respectively. Due to the higher inclination angle of GLONASS satellites, they have the most extensive latitude coverage. More satellites at high-latitude areas will be visible by this extensive latitude coverage. The ground tracks of two Galileo FOC satellites in eccentric orbit (FOC-1 and FOC-2), which are represented by the thick lines, show notable asymmetric shape. For the two Galileo satellites, the orbital inclination is 5° lower than nominal. As a result, the peak values of the latitude coverage are reduced. Nevertheless, the scientific, geodetic, and surveying applications can still use the FOC-1 and FOC-2 satellites. The ground tracks of the BDS IGSO satellites are restricted from about 76.2°E to 138.0°E longitude and 57.5°S to 57.5°N latitude. Two figure-of-eight loops can be used to describe the IGSO satellite tracks. The two loops have an average longitude difference of around 30°, so as to effectively cover the western and eastern parts of China as well as the adjoining regions. The intersection points of the two loops are at the longitude of approximately 118°E and 95°E, respectively. The availability of satellites with high elevation angles can be improved by the employment of the inclined geosynchronous orbit. Thus, for users in densely populated areas, the "urban canyon" problems can be alleviated. As shown in **Figure 1**, the BDS GEO satellites have a movement within a range of 1.8°S–1.8°N latitude, but they are fixed in longitude. The south-north movement can be attributed to the non-zero inclination. To ensure enough visible satellites for users in Asian-Pacific regions, five GEO satellites are distributed in the Indian and Pacific oceans over the equator as supplements for the IGSO satellites [5].

2.2 Navigation signals

The earlier navigation satellites provide signals on two frequencies so that the users can form dual-frequency observation combination to remove first-order

Frequency (MHz)	GPS	GLONASS	Galileo	BDS-2	BDS-3
1602 + k × 9/16		G1			
1575.42	L1		E1		B1C
1561.098				B1	B1
1278.75			E6		
1268.52				B3	B3
1246 + k × 7/16		G2			
1227.60	L2				
1207.14			E5B	B2	B2b
1202.025		G3			
1191.795			E5(A + B)		B2a + B2b
1176.45	L5		E5A		B2a

Table 1.
Navigation signals of quad-constellations.

effects of ionospheric delay based on the dispersive nature of the ionosphere. The modernized GNSS satellites have the capability of transmitting multi-frequency signals. **Table 1** details the navigation signals of the four GNSS constellations [6]. k denotes the frequency factor of GLONASS satellites. The GPS Block IIF satellites can provide signals on L1, L2, and L5 frequencies, while the other GPS satellites are still transmitting L1/L2 signals. The last seven satellites of GLONASS-M series and all GLONASS-K satellites operate with three frequency bands, namely G1, G2, and G3, while the other GLONASS satellites can only offer G1 and G2 signals. All Galileo satellites are able to broadcast E1, E5A, E5B, E5 (A + B), and E6 signals. The BDS-2 satellites are capable of providing B1, B2, and B3 triple-frequency signals. In addition to the B1 and B3 signals, the BDS-3 satellites can transmit four new navigation signals, namely, B1C, B2a, B2b, and B2a + B2b [6].

3. Kinematic single point positioning with quad-constellations

With the use of single-frequency code observations and broadcast ephemeris, the SPP technology can provide meter-level positioning accuracy. Many researchers have focused on error mitigations to improve the SPP performance. The emerging multi-GNSS integration opens new prospects. In this section, the quad-constellation integrated SPP (QISPP) model with GPS, GLONASS, BDS, and Galileo measurements is developed, and its performance in the kinematic mode is evaluated.

3.1 QISPP model

Alignment of the coordinate and time references of the four GNSS systems is a key issue for the QISPP. With respect to the coordinate references, the coordinate systems of Galileo, BDS, GLONASS, and GPS satellites adopt the broadcast orbits of GTRF, CGCS2000, PZ90.11, and WGS-84, respectively. Although different coordinate references are employed for the four GNSS systems, the differences among them are only at a level of several centimeters [7, 8]. In view that the code-based positioning solutions using broadcast ephemeris can only achieve an accuracy at a meter level, such a small difference is negligible. In other words, the four GNSS systems can directly use their satellite coordinates without coordinate transformations in the QISPP. On the other hand, it is not the case for the time scales employed by the four GNSS systems. The GPS Master Control Station establishes the GPS time, which refers to the US Naval Observatory (USNO) Coordinated Universal Time (UTC) with a small difference of <1 μs. Besides, the UTC (USNO) is periodically corrected with integer leap seconds, and thus GPS time differs from it [1]. An atomic time scale UTC (Soviet Union, SU), which is maintained by Russia with an integer difference of 3 h and a fractional difference of <1 ms, is adopted by the GLONASS system [2]. Therefore, in addition to a tiny fractional difference, the GPS time differs from the GLONASS time by leap seconds. The BDS time system (BDT) was synchronized with UTC within 100 ns at 00:00:00 on 1 January 2006, and there exists a constant offset of 14 s between the GPS time and BDT [3]. Apart from a difference of 10 nanoseconds, the Galileo System Time (GST) is nearly identical to the GPS time [4]. The differences among the time references of the four GNSS systems will significantly affect the positioning solutions. Thus, unlike the coordinate reference frames, the inconsistent time scales cannot be ignored and must be properly handled in the QISPP.

Although there is only a physical clock in the multi-GNSS receiver, receiver clock parameters with respect to their respective time scales have to be estimated for each satellite system, since different time scales are adopted by the four GNSS systems. Alternatively, instead of adding a receiver clock parameter, a system time difference parameter with respect to a reference time scale can also be introduced [9]. We can directly estimate the GPS receiver clock offset as an unknown parameter, and the receiver clock offsets of the other satellite systems are regarded as a sum of the system time difference parameter and GPS receiver clock, provided that the GPS time scale is chosen as the reference. Following Pan et al. [10], the QISPP observation model reads

$$P^g = \rho^g + cdt^g - cdT^g + d^g_{orb} + d^g_{trop} + d^g_{ion} + \varepsilon^g_P \tag{1}$$

$$P^r = \rho^r + cdt^g + cdt^{r,g}_{sys} - cdT^r + d^r_{orb} + d^r_{trop} + d^r_{ion} + \varepsilon^r_P \tag{2}$$

$$P^b = \rho^b + cdt^g + cdt^{b,g}_{sys} - cdT^b + d^b_{orb} + d^b_{trop} + d^b_{ion} + \varepsilon^b_P \tag{3}$$

$$P^e = \rho^e + cdt^g + cdt^{e,g}_{sys} - cdT^e + d^e_{orb} + d^e_{trop} + d^e_{ion} + \varepsilon^e_P \tag{4}$$

where the superscripts e, b, r, and g refer to Galileo, BDS, GLONASS, and GPS satellites, respectively. P is the measured pseudorange; ρ is the geometric range; c is the speed of light in vacuum; dt^g is the GPS receiver clock offset; $dt^{r,g}_{sys}$, $dt^{b,g}_{sys}$ and $dt^{e,g}_{sys}$ are the GPS-GLONASS, GPS-BDS, and GPS-Galileo system time differences, respectively; dT is the satellite clock offset; d_{orb} is the satellite orbit error; d_{trop} is the tropospheric delay; d_{ion} is the ionospheric delay; and ε is the measurement noise including multipath.

The broadcast ephemeris data is used to compute the clock offset and satellite position, as given in Eqs. (1)–(4). The Saastamoinen model is used to correct the tropospheric delay errors [11]. For the GPS, GLONASS, and BDS systems, the Klobuchar model is used to correct the ionospheric delay errors [12]. The second version of the NeQuick model is employed to perform the ionospheric error correction for the Galileo observations [13]. Regarding the Galileo ionospheric error corrections, the NeQuick model is better suited than the Klobuchar model [14]. Therefore, in the QISPP model, the unknown parameters to be estimated include three system time differences, one GPS receiver clock offset, and three receiver coordinates. Due to the low cost of single-frequency receivers, most SPP users are using them. Hence, in this study, we only use the single-frequency code observations of each system on the L1/G1/B1/E1 frequencies. The broadcast clock offsets and satellite orbits are referred to the ionosphere-free code combination on two frequencies. Consequently, the hardware delay biases in a form of ionosphere-free combination are contained in the satellite clock offsets derived from the broadcast ephemeris. When the ionosphere-free combined code observables are used, the hardware delay biases can be canceled out for dual-frequency users. But for single-frequency users, the hardware delay biases must be corrected. Fortunately, the broadcast navigation messages on a satellite-by-satellite basis have provided the time group delays, which can be employed to carry out the hardware delay bias corrections in the single-frequency pseudorange-based positioning.

Following Hoque et al. [15], the Klobuchar model for ionospheric error corrections of single-frequency users can be described as

$$d_{ion} = \begin{cases} F \times \left[5.0 \times 10^{-9} \times AMP \times \left(1 - \frac{x^2}{2} + \frac{x^4}{24}\right)\right] & |x| < 1.57 \\ F \times \left(5.0 \times 10^{-9}\right) & |x| \geq 1.57 \end{cases} \tag{5}$$

$$x = \frac{2\pi(t - 50400)}{PER} \tag{6}$$

$$AMP = \begin{cases} \sum_{n=0}^{3} \alpha_n \phi_m^n & AMP \geq 0 \\ 0 & AMP < 0 \end{cases} \tag{7}$$

$$PER = \begin{cases} \sum_{n=0}^{3} \beta_n \phi_m^n & PER \geq 72,000 \\ 72,000 & PER < 72,000 \end{cases} \tag{8}$$

where α and β are coefficients included as part of satellite message, ϕ_m is the geomagnetic latitude, t is the local time, and E is the elevation angle. For more details, refer to Klobuchar [12].

The second version of the NeQuick model can be depicted by several main equations provided below:

$$N_{\text{bot}}(h) = N_{\text{E}}(h) + N_{\text{F1}}(h) + N_{\text{F2}}(h) \tag{9}$$

$$N_{\text{E}}(h) = \frac{4NmE}{\left(1 + \exp\left(\frac{h - hmE}{BE}\xi(h)\right)\right)^2} \times \exp\left(\frac{h - hmE}{BE}\xi(h)\right) \tag{10}$$

$$N_{\text{F1}}(h) = \frac{4NmF1}{\left(1 + \exp\left(\frac{h - hmF1}{B1}\xi(h)\right)\right)^2} \times \exp\left(\frac{h - hmF1}{B1}\xi(h)\right) \tag{11}$$

$$N_{\text{F2}}(h) = \frac{4NmF2}{\left(1 + \exp\left(\frac{h - hmF2}{B2}\right)\right)^2} \times \exp\left(\frac{h - hmF2}{B2}\right) \tag{12}$$

$$N(h) = \frac{4NmF2}{(1 + \exp(z))^2} \exp(z) \tag{13}$$

For more details and specific meaning of the above parameters, refer to Nava et al. [13].

As to the stochastic model for the QISPP, we can use the following covariance matrix of observations:

$$Cov = \begin{bmatrix} Q_g & 0 & 0 & 0 \\ 0 & Q_r & 0 & 0 \\ 0 & 0 & Q_b & 0 \\ 0 & 0 & 0 & Q_e \end{bmatrix} \tag{14}$$

$$Q = \begin{bmatrix} \sigma_1^2 & 0 & \cdots & 0 \\ 0 & \sigma_2^2 & \cdots & 0 \\ \vdots & \vdots & \ddots & \vdots \\ 0 & 0 & \cdots & \sigma_n^2 \end{bmatrix} \tag{15}$$

where the subscript n denotes the number of satellites for each satellite system and σ^2 denotes the variance of code observations, which can be written as

$$\sigma^2 = \sigma_0^2/(\sin E)^2 \tag{16}$$

where σ_0 denotes the standard deviation (STD) of code observations, which differs among different satellite systems, and E denotes the satellite elevation angle.

The code observation precision is set to 0.3 m for GPS satellites [16]. Following Cai et al. [17], an initial weight ratio of 1:1 is appropriate for BDS and GPS code observations. Thus, the code observation precision is also set to 0.3 m for BDS satellites. The code observation precision is set to 0.6 m for GLONASS satellites due to their twice lower code chipping rate than the GPS code observations. The code observations of Galileo satellites are down-weighted by a factor of four, considering that the broadcast ephemeris has relatively lower accuracies [8]. That is, the precision of the Galileo code observations is also set to 0.6 m.

3.2 Performance analysis of QISPP solutions

Four different constellation combinations are employed for the purpose of comparison, namely, GPS/GLONASS/BDS/Galileo, GPS/GLONASS/BDS, GPS/GLONASS, and GPS-only. In the data processing, the receiver coordinates as well as other unknown parameters are estimated epoch-by-epoch without imposing any constraints between the epochs in order to analyze the single-epoch SPP performance. For brevity, in the following figures and tables, "GLO" and "GAL" are used to represent GLONASS and Galileo systems, respectively.

Figure 2 shows the epoch-wise SPP positioning errors for the four different combination cases at station NNOR on 8 April 2015. In the east, north, and up directions, the variations of positioning errors are consistent for the four cases, but the series of position errors show less fluctuation for the triple- and quad-constellation cases. It is seen that the GPS-only SPP achieves larger positioning errors than the GPS/GLONASS case. The positioning errors of GPS/GLONASS/BDS SPP at almost all epochs are further reduced by combining with BDS. The further introduction of Galileo observations does not exhibit significant change, since the blue lines are almost completely covered by the orange ones.

Figure 3 presents the position dilution of precision (PDOP) and the number of visible satellites for the four cases. It is clear that the number of visible satellites is obviously increased by the multi-constellation combination and the PDOP value is simultaneously decreased. The quad-constellation integrated case increases the average number of visible satellites from 6.9 to 27.8 in comparison to the GPS-only SPP, and thus the average PDOP values are significantly decreased from 2.3 to 1.1. The increased number of available satellites and decreased PDOP values explain

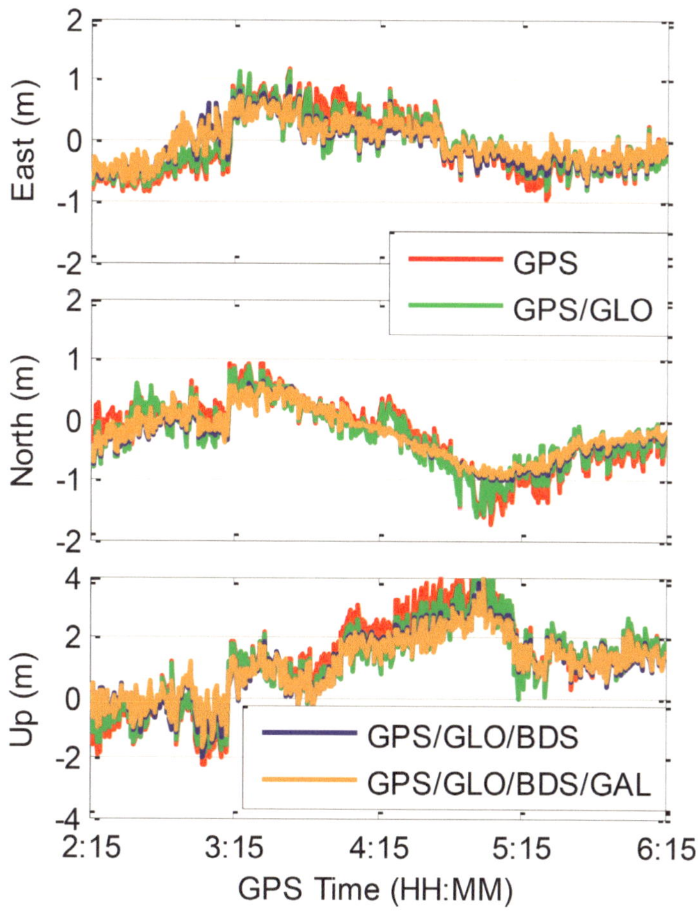

Figure 2.
Kinematic positioning errors of four different combination cases for the SPP processing at NNOR.

why the positioning accuracies can be improved in the multi-constellation integrated cases.

The datasets from 47 International GNSS Service (IGS) stations are processed for further analysis, and the average root mean square (RMS) statistics of epoch-wise SPP positioning errors, number of satellites, and PDOP are obtained. The average satellite numbers of the above four processing cases are 8.6, 15.6, 21.1, and 23.4, respectively, while the corresponding average PDOPs are 2.0, 1.4, 1.2, and 1.1, respectively. The combination of GPS and GLONASS improves the positioning accuracy over the GPS-only case by 7, 5, and 5% from 0.81, 2.05, and 3.13 m to 0.75, 1.94, and 2.96 m in the east, north, and up directions, respectively. In the triple-constellation SPP, an accuracy improvement of 9, 6, and 7% over the GPS/GLONASS case to 0.68, 1.82, and 2.75 m in the three directions is achieved, respectively. After a further integration with Galileo, the positioning accuracy is only improved by 3, 2, and 2% to 0.66, 1.78, and 2.70 m in the three directions, respectively [10].

4. Kinematic precise point positioning with quad-constellations

The PPP technique adopts an absolute positioning approach to achieve centimeter-level positioning accuracy using code and carrier phase observations as well as precise satellite orbit and clock offset corrections [18]. Both industrial

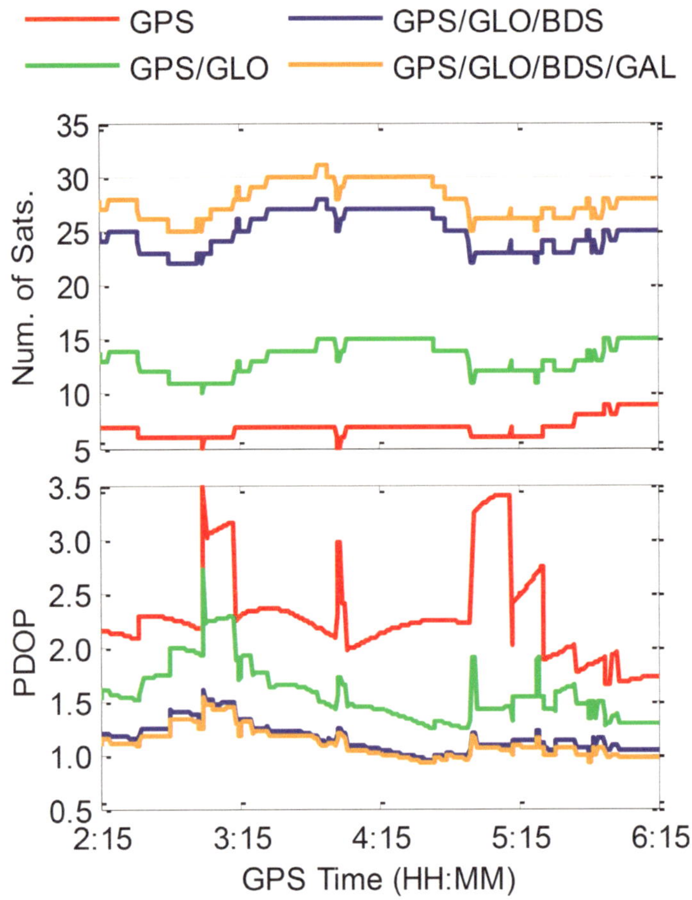

Figure 3.
Number of satellites and PDOP values for the SPP processing at NNOR.

applications and scientific research widely use GPS-based PPP. But the position solutions of GPS-based PPP require a long time to converge. Due to high measurement redundancy, significantly reduced convergence time and improved positioning accuracy can be expected by using multi-constellation GNSS PPP. The positioning model and processing method of quad-constellation integrated PPP (QIPPP) with GPS, GLONASS, BDS, and Galileo measurements are developed, and then the improvement in precision and convergence time from quad-constellations is verified by comparing the solutions of different constellation combinations.

4.1 QIPPP model

In view that the code observation equation is detailed in Eqs. (1)–(4), only the carrier phase observations on ith (i = 1, 2) frequency are formulated, that is:

$$\Phi_i = \rho + cdt - cdT + d_{orb} + d_{trop} - d_{ion/L_i} + \lambda_i N_i + \varepsilon_{\Phi_i} \tag{17}$$

where Φ is the measured carrier phase, $d_{ion/Li}$ is the ionospheric delay on ith frequency, λ is the wavelength, and N is the phase ambiguity term grouped with hardware delays. It should be noted that the wavelength λ is different for different GLONASS satellites because GLONASS employs the frequency-division multiple access (FDMA) technique.

To remove the first-order effects of ionospheric delays, the PPP normally utilizes the ionosphere-free (IF) combined observables, that is:

$$P_{IF} = \left(f_1^2 \cdot P_1 - f_2^2 \cdot P_2\right) / \left(f_1^2 - f_2^2\right) \tag{18}$$

$$\Phi_{IF} = \left(f_1^2 \cdot \Phi_1 - f_2^2 \cdot \Phi_2\right) / \left(f_1^2 - f_2^2\right) \tag{19}$$

where Φ_{IF} and P_{IF} denote the IF combined carrier phase and code observables, respectively, and f denotes the carrier phase frequency, which differs among GLONASS satellites.

In order to investigate emerging new satellite systems such as Galileo and BDS, the Multi-GNSS Experiment (MGEX) has been established by the IGS [19]. The correction of satellite clock and orbit errors for QIPPP is conducted with the use of precise satellite clock and orbit products provided by the MGEX. The first-order effects of ionospheric delays are removed using the IF combined observables, as shown in Eqs. (18) and (19). The tropospheric delays can be divided into a wet part and a dry part [20]. The wet part is estimated from the measurements, while the Hopfield tropospheric model is employed to correct the dry part. The projection from slant delays to zenith delays adopts the Niell mapping functions [21]. As some literatures such as Kouba and Héroux [18] have well described other error mitigations, they are not provided here. In PPP, the code-specific hardware delays at the receiver and the receiver clock offsets are usually estimated as a lumped term, as they are linearly correlated with each other. For different navigation systems, both the frequency and signal structures differ. Consequently, within a multi-GNSS receiver, the receiver-dependent code hardware delays are different for the four navigation systems. For the purpose of solving this issue, we should design a receiver clock offset parameter for each satellite system. Alternatively, the differences between receiver clock estimates of different satellite systems can be compensated by introducing an inter-system bias (ISB). The QIPPP observation model can be written as follows, provided that the GPS system is chosen as the reference [22]:

$$P_{IF}^{g} = \rho^{g} + cdt + M^{g} d_{zwd} + \varepsilon_{P_{IF}}^{g} \tag{20}$$

$$\Phi_{IF}^{g} = \rho^{g} + cdt + M^{g} d_{zwd} + N_{IF}^{g} + \varepsilon_{\Phi_{IF}}^{g} \tag{21}$$

$$P_{IF}^{b} = \rho^{b} + cdt + \mathrm{ISB}_{b,g} + M^{b} d_{zwd} + \varepsilon_{P_{IF}}^{b} \tag{22}$$

$$\Phi_{IF}^{b} = \rho^{b} + cdt + \mathrm{ISB}_{b,g} + M^{b} d_{zwd} + N_{IF}^{b} + \varepsilon_{\Phi_{IF}}^{b} \tag{23}$$

$$P_{IF}^{r} = \rho^{r} + cdt + \mathrm{ISB}_{r,g} + M^{r} d_{zwd} + \varepsilon_{P_{IF}}^{r} \tag{24}$$

$$\Phi_{IF}^{r} = \rho^{r} + cdt + \mathrm{ISB}_{r,g} + M^{r} d_{zwd} + N_{IF}^{r} + \varepsilon_{\Phi_{IF}}^{r} \tag{25}$$

$$P_{IF}^{e} = \rho^{e} + cdt + \mathrm{ISB}_{e,g} + M^{e} d_{zwd} + \varepsilon_{P_{IF}}^{e} \tag{26}$$

$$\Phi_{IF}^{e} = \rho^{e} + cdt + \mathrm{ISB}_{e,g} + M^{e} d_{zwd} + N_{IF}^{e} + \varepsilon_{\Phi_{IF}}^{e} \tag{27}$$

where $\mathrm{ISB}_{b,g}$, $\mathrm{ISB}_{r,g}$, and $\mathrm{ISB}_{e,g}$ are the GPS-BDS, GPS-GLONASS, and GPS-Galileo inter-system biases, respectively, d_{zwd} is the tropospheric zenith wet delay (ZWD), and M is the tropospheric mapping function.

For the QIPPP processing, we employ a Kalman filter approach. Actually, the geometric range ρ in Eqs. (20)–(27) is a function of receiver coordinates and satellite coordinates. After the precise ephemeris data is used to determine the satellite coordinates and the ρ is linearized, the unknown parameters include phase ambiguity parameters equal to the number of the observed GNSS satellites, one

ZWD, three ISB, one receiver clock offset, and three receiver coordinates. We should also provide appropriate dynamic models for the state vector and proper stochastic models for the measurements in the Kalman filter. The IF combined phase and code observables are obtained using the raw phase and code measurements on two different frequencies in a form of linear combination. It is assumed that the measurements on different frequencies are independent. Based on the law of random error propagation [23], we can derive the initial variances of the IF combined observables. The satellite elevation angles can be further introduced to acquire the actual variances [24]. As for the dynamic models, the phase ambiguity parameters can be modeled as constants, whereas the ZWD, ISB, receiver clock offset, and kinematic receiver coordinates may be modeled as a random walk (RW) process [25–27].

4.2 Performance analysis of QIPPP solutions

The datasets from stations SEYG, JFNG, and MAR7 on 1 March 2019 are used for numerical analysis in this section. The three stations are located at low-, middle-, and high-latitude regions, respectively, and all of them are able to provide multi-constellation observations. The in-house MIPS-PPP software capable of processing quad-system observation data as well as single-system measurements of Galileo, BDS, GLONASS, and GPS, which is developed at Central South University, China, is employed for the quad-constellation GNSS PPP processing. Regarding the specific PPP position determination, the Galileo E1/E5A, BDS B1/B2, GLONASS G1/G2, and GPS L1/L2 dual-frequency observations are used. The cutoff satellite angle is set to 10°, while the observations are recorded at a sampling rate of 30 s. The IGS Analysis Center, German Research Centre for Geosciences (GFZ), has been generating and releasing the mixed multi-GNSS final precise satellite clock offset and orbit products with a sampling interval of 30 s and 5 min, respectively, and the mitigation of the satellite clock and orbit errors is carried out using them in this study. For the ISB, receiver clock offset, and ZWD parameters, the spectral density values are set to 10^{-7}, 10^{5}, and 10^{-9} m^2/s, respectively [28]. The initial STD values for GPS and GLONASS code observations are set to 0.3 and 0.6 m, while for phase observations they are both set to 2 mm [28]. Since there is a relatively lower accuracy for the satellite orbits and clocks of Galileo and BDS [29, 30], their observations are down-weighted with a factor of four. That is, for both Galileo and BDS, the code and phase observation accuracies are set to 0.6 m and 4 mm, respectively. Kinematic processing is made on an epoch-by-epoch basis using the static data. No constraints between epochs are imposed so as to simulate kinematic situations. In the Kalman filtering, the coordinates of the dynamic receiver are modeled as a RW process, and the spectral density is set as 10^2 m^2/s.

Figure 4 shows the positioning errors of GPS, GPS/BDS, GPS/BDS/GLONASS, and four-system PPP in three directions of east, north, and up in the simulated kinematic test. It can be seen from **Figure 4** that, compared with the PPP solutions of GPS-only system, the error curve of the PPP of multi-constellation combinations converges to the stable value faster in the east, north, and up directions. For all processing schemes, the positioning errors in the vertical direction are larger than those of horizontal directions.

In order to assess the kinematic positioning accuracy, **Table 2** provides the RMS statistical values using the position errors in the last 1 h in which the position solutions in all three components have reached stable values. The results show that, taking JFNG station as an example, the positioning accuracy of the GPS-only PPP in three directions is 0.160, 0.097, and 0.192 m, respectively. After the combination of GPS and BDS, compared with the single GPS system, the positioning accuracy is

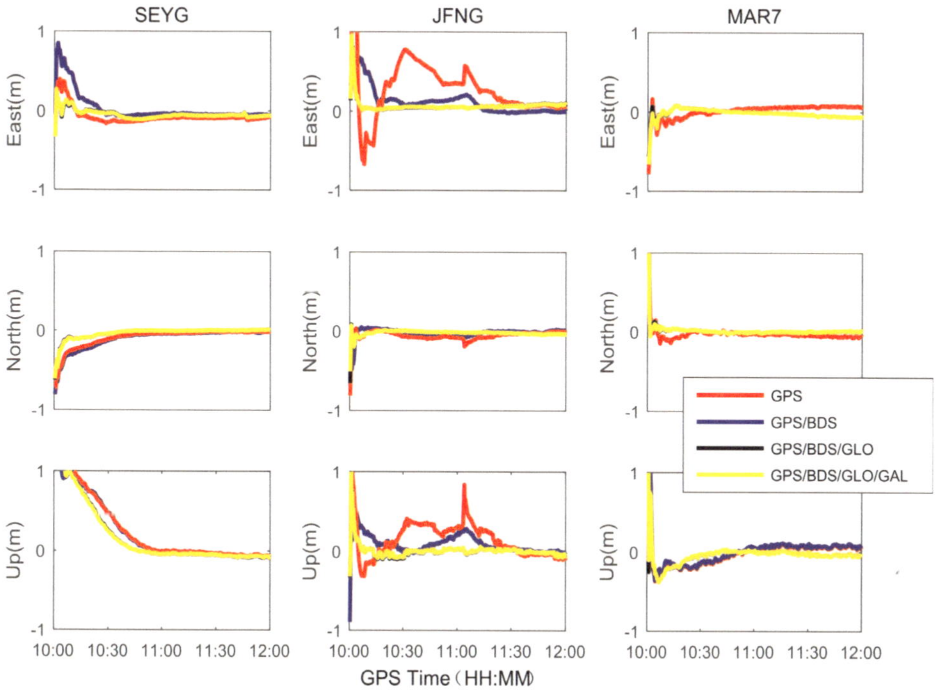

Figure 4.
PPP kinematic positioning errors at stations SEYG, JFNG, and MAR7 for four different processing cases.

Stations	Directions	GPS	GPS/BDS	GPS/BDS/GLO	GPS/BDS/GLO/GAL
SEYG	East	0.090	0.061	0.058	0.043
	North	0.062	0.057	0.028	0.023
	Up	0.158	0.126	0.082	0.071
	3D	0.192	0.151	0.104	0.086
JFNG	East	0.160	0.142	0.052	0.050
	North	0.097	0.044	0.020	0.020
	Up	0.192	0.158	0.119	0.106
	3D	0.268	0.217	0.131	0.119
MAR7	East	0.066	0.054	0.040	0.032
	North	0.045	0.035	0.028	0.021
	Up	0.081	0.068	0.034	0.030
	3D	0.114	0.094	0.060	0.049

Table 2.
Convergence accuracy of kinematic PPP (unit: m).

improved by 11, 55, and 18%, respectively. Compared with the GPS/BDS PPP, the positioning accuracy of the combination of three systems is significantly improved. After further adding Galileo observations, the three-dimensional (3D) accuracy of PPP of the four-system combination is slightly improved by 1.2 cm. **Table 3** shows the convergence time in three directions. Taking the JFNG station as an example, the PPP solutions of the four-system combination requires 22.5, 28.5, and 79.0 min to converge to the accuracy level of 1 dm. Compared with the single- and

Stations	Directions	GPS	GPS/BDS	GPS/BDS/GLO	GPS/BDS/GLO/GAL
SEYG	East	70.5	64.0	51.5	51.5
	North	37.5	33.0	15.0	14.5
	Up	90.0	88.0	77.0	76.5
JFNG	East	94.0	85.0	23.0	22.5
	North	86.0	79.5	28.5	28.5
	Up	88.0	82.5	79.0	79.0
MAR7	East	79.0	78.0	68.5	68.5
	North	43.0	41.0	23.5	23.5
	Up	86.0	84.5	61.0	60.0

Table 3.
Convergence time of kinematic PPP (unit: min).

dual-system cases, the convergence time in the horizontal directions of the triple- and quad-system integrated PPP is dramatically shortened. In view of the fact that the positioning accuracy of kinematic PPP in the vertical direction is worse than that of the horizontal directions, the convergence standard may be too strict for the vertical direction, which even leads to the failure of effective convergence in a short period of time sometimes.

5. Conclusions

The GNSS-based absolute positioning technologies can provide reliable kinematic position services anywhere, in all weather, and anytime using a single receiver. With single-frequency code measurements and broadcast satellite ephemeris, the SPP technology can provide meter-level positioning accuracy. With dual-frequency code and carrier phase measurements as well as precise satellite orbit and clock products, the PPP technology can offer centimeter-level positioning accuracy. In recent years, the satellite systems have been booming. In view that both SPP and PPP belong to the satellite-based kinematic absolute positioning technologies, the multi-constellation combination provides new prospects for their performance improvement, due to more visible satellites, increased measurement redundancy, and enhanced satellite sky distribution. The quad-constellation integrated SPP and PPP models with GPS, GLONASS, BDS, and Galileo measurements are developed, respectively. The results indicate significantly improved positioning performance of the multi-GNSS integration, which will further promote the applications of SPP and PPP technologies.

Acknowledgements

The contribution of data and products from IGS is appreciated.

Conflict of interest

The authors declare no conflict of interest.

List of abbreviations

BDS	BeiDou Navigation Satellite System
BDS-1	BeiDou navigation demonstration system
BDS-2	regional BDS
BDS-3	global BDS
BDS-3S	BDS-3 demonstration system
BDT	BDS time system
FDMA	frequency-division multiple access
FOC	full operational capability
GEO	geostationary earth orbit
GFZ	German Research Centre for Geosciences
GLONASS	Global Navigation Satellite System
GNSS	Global Navigation Satellite System
GST	Galileo System Time
IF	ionosphere-free
IGS	International GNSS Service
IGSO	inclined geosynchronous orbit
IOV	in-orbit validation
ISB	inter-system bias
MEO	medium earth orbit
MGEX	multi-GNSS experiment
PDOP	position dilution of precision
PNT	positioning, navigation, and timing
PPP	precise point positioning
QIPPP	quad-constellation integrated PPP
QISPP	quad-constellation integrated SPP
RMS	root mean square
RW	random walk
SPP	single point positioning
STD	standard deviation
SU	Soviet Union
USNO	United States Naval Observatory
UTC	Coordinated Universal Time
ZWD	zenith wet delay
3D	three-dimensional

Author details

Lin Pan, Changsheng Cai*, Jianjun Zhu and Xianqiang Cui
School of Geosciences and Info-Physics, Central South University, Changsha, China

*Address all correspondence to: cscai@hotmail.com

References

[1] GPS Directorate. Navstar GPS space segment navigation user segment interfaces, Interface specification (IS-GPS-200). Revision G. Washington: Global Positioning System Directorate; 2012

[2] RISDE. Global Navigation Satellite System GLONASS Interface Control Document. Version 5.1. Moscow: Russian Institute of Space Device Engineering; 2008

[3] CSNO. BeiDou Navigation Satellite System Signal in Space Interface Control Document (Open Service Signal). Version 2.0. Beijing: China Satellite Navigation Office; 2013

[4] EU. European GNSS (Galileo) open service signal in space interface control document (OS-SIS-ICD). Issue 1.1. European Union; 2010

[5] Pan L, Zhang X, Li X, Li X, Lu C, Liu J, et al. Satellite availability and point positioning accuracy evaluation on a global scale for integration of GPS, GLONASS, BeiDou and Galileo. Advances in Space Research. 2019; **63**(9):2696-2710. DOI: 10.1016/j.asr. 2017.07.029

[6] RTCM-SC104. The Receiver Independent Exchange Format (RINEX). Version 3.04. International GNSS Service (IGS), RINEX Working Group and Radio Technical Commission for Maritime Services Special Committee 104 (RTCM-SC104). 2018

[7] Torre AD, Caporali A. An analysis of intersystem biases for multi-GNSS positioning. GPS Solutions. 2015;**19**(2): 297-307

[8] Montenbruck O, Steigenberger P, Hauschild A. Broadcast versus precise ephemerides: A multi-GNSS perspective. GPS Solutions. 2015;**19**(2): 321-333

[9] Cai C, Gao Y. A combined GPS/ GLONASS navigation algorithm for use with limited satellite visibility. Journal of Navigation. 2009;**62**(4): 671-685

[10] Pan L, Cai C, Santerre R, Zhang X. Performance evaluation of single-frequency point positioning with GPS, GLONASS, BeiDou and Galileo. Survey Review. 2017;**49**(354):197-205. DOI: 10.1080/00396265.2016.1151628

[11] Saastamoinen J. Contribution to the theory of atmospheric refraction. Bulletin Géodésique. 1973;**107**(1):13-34

[12] Klobuchar J. Ionospheric time-delay algorithms for single-frequency GPS users. IEEE Transactions on Aerospace and Electronic Systems. 1987;**AES-23** (3):325-331

[13] Nava B, Coïsson P, Radicella SM. A new version of the NeQuick ionosphere electron density model. Journal of Atmospheric and Solar-Terrestrial Physics. 2008;**70**(15):1856-1862

[14] Oladipo OA, Schüler T. GNSS single frequency ionospheric range delay corrections: NeQuick data ingestion technique. Advances in Space Research. 2012;**50**(9):1204-1212

[15] Hoque MM, Jakowski N, Berdermann J. An ionosphere broadcast model for next generation GNSS. In: Proceedings of the 28th International Technical Meeting of the ION Satellite Division, ION GNSS+ 2015; 14-18 September 2015; Tampa, Florida, USA. pp. 3755-3765

[16] Cai C, Gao Y, Pan L, Dai W. An analysis on combined GPS/COMPASS data quality and its effect on single point positioning accuracy under different observing conditions. Advances in Space Research. 2014;**54**(5):818-829

[17] Cai C, Pan L, Gao Y. A precise weighting approach with application to combined L1/B1 GPS/BeiDou positioning. Journal of Navigation. 2014; **67**(5):911-925

[18] Kouba J, Héroux P. Precise point positioning using IGS orbit and clock products. GPS Solutions. 2001;**5**(2): 12-28. DOI: 10.1007/PL00012883

[19] Rizos C, Montenbruck O, Weber R, Weber G, Neilan R, Hugentobler U. The IGS MGEX experiment as a milestone for a comprehensive multi-GNSS service. In: Proceedings of the ION 2013 Pacific PNT Meeting (ION-PNT-2013); 23-25 April 2013; Honolulu, Hawaii, USA. pp. 289-295

[20] Davis JL, Herring TA, Shapiro II, Rogers AEE, Elgered G. Geodesy by radio interferometry: Effects of atmospheric modeling errors on estimates of baseline length. Radio Science. 1985;**20**(6):1593-1607. DOI: 10.1029/RS020i006p01593

[21] Niell AE. Global mapping functions for the atmosphere delay at radio wavelengths. Journal of Geophysical Research. 1996;**101**(B2):3227-3246. DOI: 10.1029/95JB03048

[22] Cai C, Gao Y, Pan L, Zhu J. Precise point positioning with quad-constellations: GPS, BeiDou, GLONASS and Galileo. Advances in Space Research. 2015;**56**(1):133-143. DOI: 10.1016/j.asr. 2015.04.001

[23] Xu G. GPS: Theory, Algorithms and Applications. 2nd ed. Berlin: Springer; 2007

[24] Gerdan GP. A comparison of four methods of weighting double difference pseudorange measurements. The Australian Surveyor. 1995;**40**(4):60-66

[25] Axelrad P, Brown RG. GPS navigation algorithms. In: Parkinson BW, Spilker JJ, editors. Global Positioning System: Theory and Applications. Progress in Astronautics and Aeronautics. Virginia: American Institute of Astronautics and Aeronautics; 1996. pp. 409-433

[26] Brown RG, Hwang PY. Introduction to Random Signals and Applied Kalman Filtering. 3rd ed. New York: Wiley; 1997

[27] Dodson AH, Shardlow PJ, Hubbard LCM, Elgered G, Jarlemark POJ. Wet tropospheric effects on precise relative GPS height determination. Journal of Geodesy. 1996;**70**(4):188-202. DOI: 10.1007/BF00873700

[28] Cai C, Gao Y. Modeling and assessment of combined GPS/GLONASS precise point positioning. GPS Solutions. 2013;**17**(2):223-236. DOI: 10.1007/s10291-012-0273-9

[29] Zhao Q, Guo J, Li M, Qu L, Hu Z, Shi C, et al. Initial results of precise orbit and clock determination for COMPASS navigation satellite system. Journal of Geodesy. 2013;**87**(5):475-486. DOI: 10.1007/s00190-013-0622-7

[30] Steigenberger P, Hugentobler U, Loyer S, Perosanz F, Prange L, Dach R, et al. Galileo orbit and clock quality of the IGS multi-GNSS experiment. Advances in Space Research. 2015;**55**(1): 269-281. DOI: 10.1016/j.asr.2014.06.030

Chapter 6

Kinematics for Spacecraft-Type Robotic Manipulators

Ijar Milagre da Fonseca, Maurício Nacib Pontuschka and Glaydson Luiz Bertoze Lima

Abstract

The scope of this chapter is the study of the forward and inverse kinematics for a space robot. The main focus is to compute the position and orientation of manipulators' end-effectors relative to their platform. Such platform plays the role of workstations referred in the literature approaching ground manipulators. In this study, the method is to write the manipulator kinematics' equations as functions of the joint variables by following the Denavit-Hartenberg convention. The homogeneous transform technique is used to study the kinematics. The set of coordinate frames defined in this chapter follows the convention for frames that appears in the literature for ground robot manipulators. The kinematics related to the spacecraft attitude is added in the formulation because the manipulator studied in this chapter is type spacecraft. The objective is to provide an overview and clear understanding of the kinematics' equations for spacecraft-type manipulators. To be consistent with orbital dynamics area, the inertial, orbital, and body-fixed coordinate frames are included in this kinematics study. The forward and inverse kinematics formulations are derived. The MATLAB®/Simulink tools are presented for the computer simulations of the forward and inverse kinematics.

Keywords: space robot manipulators, forward kinematics, inverse kinematics, attitude, orbit

1. Introduction

Kinematics is mainly concerned with the geometry of motion. The subject of kinematics is to some extent mathematical and does not consider any forces associated with the motion. A comprehensive study of kinematics is a prerequisite to the successful formulation of the equations of motion and the dynamic analysis. Consistent with the geometry of motion, kinematics involves the definition of systems of reference, methods of establishing relationships between frames, and algebraic manipulations of matrices and vectors. Kinematics is primarily concerned with describing the orientation of a body with respect to a known reference or coordinate frame. When dealing with robotic manipulators, we have to include also translational kinematics. In this chapter, the calculation goes beyond the algebraic manipulations of rotation matrices connecting frames. This includes the derivation of the equations for the rotational and the translational kinematics that characterize robot manipulators' kinematics. Space robotic manipulators are designed to implement orbital operations. There is a considerable difference regarding Earth-based manipulators, as they increase the type and number of reference systems for proper problem formulation [1–4]. The main differences with regard to Earth-based robotic manipulators are the inclusion of an Earth-centered inertial coordinate frame (ECI), besides the local vertical, local horizontal (LVLH) reference frame, and the spacecraft fixed reference frame. These frames are used to describe the kinematics' differential equations related to orbit and attitude motions. The other reference frames are typical of those defined for ground manipulators. For space manipulators, we need to consider the relationships between all these reference systems.

2. Forward kinematics of manipulators and D-H convention—homogeneous transforms

Forward kinematics is the problem of computing the position and orientation of the robot tool with respect to a fixed reference frame defined on the base of the space robot. Such manipulators can be seen as a set of links connected in a chain of joints. The joints are points of connection between a pair of links. The joints of a manipulator may be revolute (rotatory), prismatic (sliding), or a combination of both. According to that link to joint configuration, any robot manipulator kinematics can be described in terms of four quantities associated with each link. Two of these quantities describe the link itself. The other two describe the link's connection to the neighboring link. For revolute joints, the angle θ_i is the joint variable, and the other three quantities are fixed link parameters. These parameters are illustrated in **Figure 1** as the twist angle α_{i-1}, the distances a_{i-1}, and d_i. For prismatic joints, the joint variable is d_i, while the angle θ_i is zero. In this case, α_{i-1}, and a_{i-1} are fixed link parameters (see **Figure 1**). The description of manipulators by these quantities is referred as the Denavit-Hartenberg convention, denoted here as D-H [5, 6]. The understanding of those four quantities requires the frame convention shown in **Figure 2**. The frames are denoted by the symbol {}, for example {*i-1*} denotes a frame whose origin is in joint *i-1*. A summary of the link parameters considering frames convention shown in **Figure 2** can be written as:

- a_i is the distance measured along x-axis from z_i to z_{i+1}
- α_i is the angle rotation measured about x_i from z_i to z_{i+1}
- d_i is the distance measured along z_i from x_{i-1} to x_i
- θ_i is a joint angle measured about z_i from x_{i-1} to x_i.

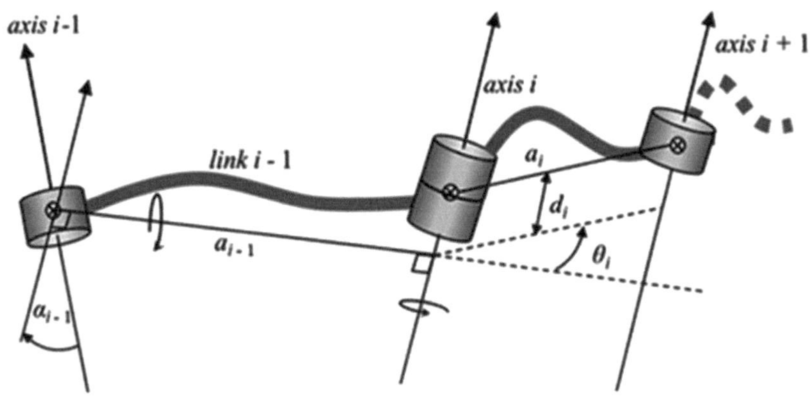

Figure 1.
D-H convention showing the notation, joint variable, and parameters.

DOI: http://dx.doi.org/10.5772/intechopen.85636

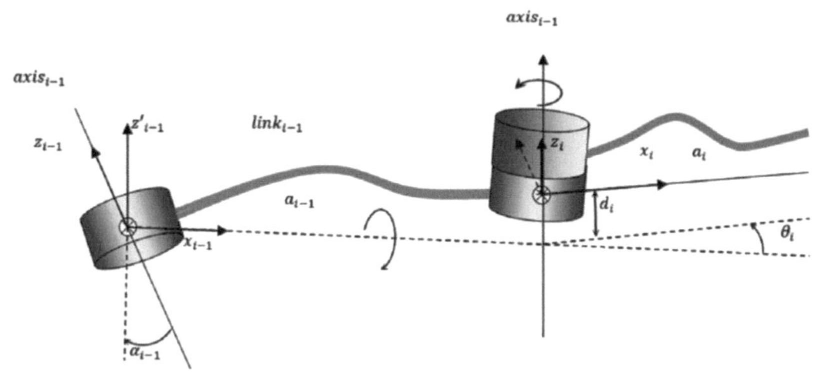

Figure 2.
Frames, links, and D-H quantities.

The D H convention describes the manipulator rotational plus translational kinematics by using homogeneous transform. This transform is a 4 × 4 matrix in the form:

$$\begin{Bmatrix} {}^{i-1}\boldsymbol{p} \\ 1 \end{Bmatrix} = \begin{bmatrix} \cos\theta_i & -\sin\theta_i & 0 & {}^{i-1}p_x \\ \sin\theta_i & \cos\theta_i & 0 & {}^{i-1}p_y \\ 0 & 0 & 1 & {}^{i-1}p_z \\ 0 & 0 & 0 & 1 \end{bmatrix} \begin{Bmatrix} {}^{i}\boldsymbol{p_o} \\ 1 \end{Bmatrix} = \begin{bmatrix} {}^{1-1}_{\ \ i}\boldsymbol{R} & {}^{i-1}\boldsymbol{p_o} \\ \boldsymbol{0} & \boldsymbol{1} \end{bmatrix} \begin{Bmatrix} {}^{i}\boldsymbol{p} \\ 1 \end{Bmatrix} \quad (1)$$

The result is ${}^{i-1}_{\ \ i}\boldsymbol{p} = {}^{i-1}_{\ \ i}\boldsymbol{R}^{i}\boldsymbol{p} + {}^{i-1}\boldsymbol{p_o}$ and **1** = **1.** The subscript o is just to say that the position of the end-effector is given with respect to the origin of the frame ${}^{1-1}_{\ \ i}\boldsymbol{R}$.

The 3 × 3 ${}^{1-1}_{\ \ i}\boldsymbol{R}$ block matrix is related to the rotation between frames $\{i-1\}$ and $\{i\}$ by the joint angle θ_i. The 3 × 1 vector ${}^{i-1}\boldsymbol{p_o}$ is related to the end-effector position with respect o the frame $\{i-1\}$. The advantage of this approach is the use of matrix notation including the position in addition to the orientation.

Such matrix notation facilitates the mathematical formulation in a compact and comprehensive format. Also, the D-H formulation is appropriate to implement problems in the MATLAB® computational environment.

Let us discuss now the manipulation of the frames as defined in **Figure 2** and their parameters. The D-H convention considers that all the joint angle rotations take place along z-axis. The rule is the same for the joint variables d_i (prismatic joints). In the case of prismatic joints, the joint angle is zero. This convention regarding joint variables along z-axis means that when z-axis is not parallel to the next one, a twist angle rotation α must be done so as to generate a new z-axis parallel to the next one. Angle α can be positive or negative. Another convention applies to the distance a_i. Once we have the new z-axis parallel to the next one, the distance a_i is determined by drawing a mutual perpendicular line between both z-axes. Such line will determine the distance between the joints and is also the place of a new x-axis. A y-axis completes the right-handed frame (**Figure 2**). If the line containing a_i crosses the z-axis offset of the joint, it defines a distance d_i as shown in **Figure 2**. Finally, we rotate the z-axis by the joint variable θ_i and a new frame is generated, taking into account the offset d_i.

On the base of the space robot attachment, the $Ox_0y_0z_0$ frame or $\{0\}$ is defined. Note that the frame is $\{i-1\}$, and for i = 1, this frame is $\{0\}$. In the literature of robot manipulators, $\{0\}$ frame does not move with respect to the space robot platform. Accordingly, the joint associated to $\{0\}$ is also named joint 0 from which the link 0 connects to the joint 1. It is common to define joint 0 coincident with joint 1. In this case, a_0 is zero. For ground robots, it is common to define $Ox_0y_0z_0$ as an inertial frame. However, such definition is not consistent with the concept of inertial frames, mainly when we have space robot orbiting the Earth. For a space robotic manipulator, the $Ox_0y_0z_0$ frame is defined with its origin somewhere on the platform and locates the placement of joint zero. **Figures 2** and **3** illustrate the frames, joint axes, joint variables, and the other fixed quantities according to the D-H convention.

To include the translation in the same matrix notation, we use the homogeneous transform as shown in Eq. (1).

Before going to the other frames definition for the spacecraft, let us implement the matrix algebra, considering **Figure 2**. To clarify the procedure to obtain the general expression for the rotational-translational kinematics, let us redefine some frames in **Figure 2**.

- Let the frame formed by $x'_{i-1}, y'_{i}, z'_{i-1}$, obtained be a rotation α_{i-1} about x_{i-1}, be defined as $\{R\}$;
- Consider frame formed by $x'_{i-1}, y'_{i-1}, z'_{i-1}$ translated by a_{i-1} as $\{Q\}$;
- Consider the frame obtained by a rotation θ_i and parallel to frame $\{i\}$ as $\{P\}$;
- Consider frame $\{i\}$ obtained by a translation d_i.

Consider now **Figure 2** that shows the various frames in addition to the D-H parameters.

To write the kinematics' equations in the frame $\{i-1\}$, we need to compute the rotation matrices using the homogeneous transform as below

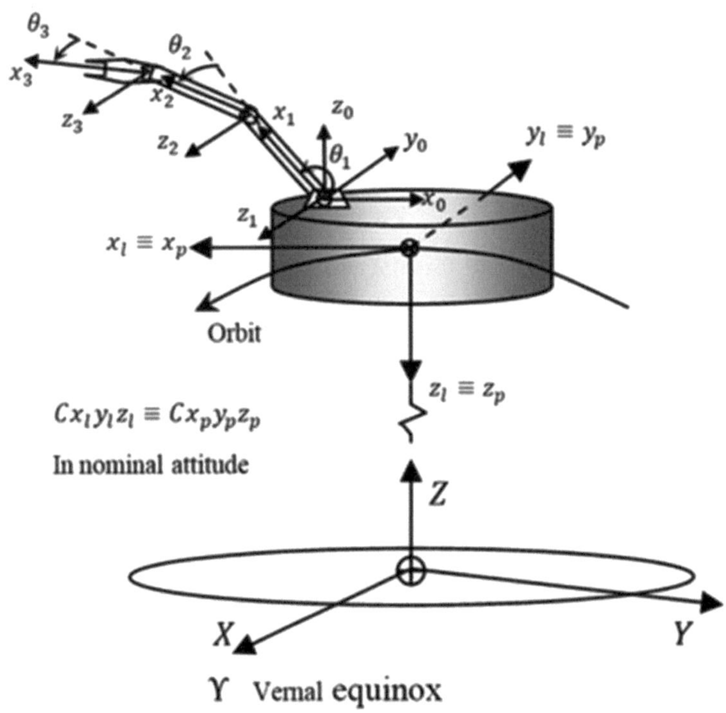

Figure 3.
Space robot on-orbit configuration and frames.

$$
\begin{aligned}
{}^{i-1}P &= {}^{i-1}_{R}T\,{}^{R}_{Q}T\,{}^{Q}_{P}T\,{}^{P}_{i}T\;{}^{i}P = {}^{i-1}_{i}T\;{}^{i}P \\
{}^{i-1}_{i}T &= \begin{bmatrix} 1 & 0 & 0 & 0 \\ 0 & c\alpha_{i-1} & -s\alpha_{i-1} & 0 \\ 0 & s\alpha_{i-1} & c\alpha_{i-1} & 0 \\ 0 & 0 & 0 & 1 \end{bmatrix} \begin{bmatrix} 1 & 0 & 0 & a_{i-1} \\ 0 & 1 & 0 & 0 \\ 0 & 0 & 1 & 0 \\ 0 & 0 & 0 & 1 \end{bmatrix} \begin{bmatrix} c\theta_i & -s\theta_i & 0 & 0 \\ s\theta_i & c\theta_i & 0 & 0 \\ 0 & 0 & 1 & 0 \\ 0 & 0 & 0 & 1 \end{bmatrix} \begin{bmatrix} 1 & 0 & 0 & 0 \\ 0 & 1 & 0 & 0 \\ 0 & 0 & 1 & d_i \\ 0 & 0 & 0 & 1 \end{bmatrix} \\
{}^{i\ 1}_{i}T &= \begin{bmatrix} 1 & 0 & 0 & a_{i-1} \\ 0 & c\alpha_{i-1} & -s\alpha_{i-1} & 0 \\ 0 & s\alpha_{i-1} & c\alpha_{i-1} & 0 \\ 0 & 0 & 0 & 1 \end{bmatrix} \begin{bmatrix} c\theta_i & -s\theta_i & 0 & 0 \\ s\theta_i & c\theta_i & 0 & 0 \\ 0 & 0 & 1 & d_i \\ 0 & 0 & 0 & 1 \end{bmatrix} \\
&= \begin{bmatrix} c\theta_i & -s\theta_i & 0 & a_{i-1} \\ s\theta_i c\alpha_{i-1} & c\theta_i c\alpha_{i-1} & -s\alpha_{i-1} & -s\alpha_{i-1} d_i \\ s\theta_i s\alpha_{i-1} & c\theta_i s\alpha_{i-1} & c\alpha_{i-1} & c\alpha_{i-1} d_i \\ 0 & 0 & 0 & 1 \end{bmatrix}
\end{aligned} \tag{2}
$$

where ${}^{i-1}_{i}T = {}^{i-1}_{R}T\,{}^{R}_{Q}T\,{}^{Q}_{P}T\,{}^{P}_{i}T$ and the subscript to superscript means transform from the previous to the next coordinate frame.

These abbreviations will be used throughout this chapter.

$$s_i = s\theta_i = \sin\theta_i \tag{3}$$

$$c_i = c\theta_i = \cos\theta_i \tag{4}$$

$$c_{12} = c(\theta_1 + \theta_2) = c_1c_2 - s_1s_2 = c\theta_1 c\theta_2 - s\theta_1 s\theta_2 \tag{5}$$

$$s_{12} = s(\theta_1 + \theta_2) = c_1s_2 + s_1c_2 = c\theta_1 s\theta_2 - s\theta_1 c\theta_2 \tag{6}$$

$$c_{123} = c\left(\sum_i^3 \theta_i\right) \text{ and } s_{123} = s\left(\sum_i^3 \theta_i\right) \text{ for } i = 1, 2, 3 \tag{7}$$

In Eq. (2), the first matrix,

$$\begin{bmatrix} 1 & 0 & 0 & 0 \\ 0 & c\alpha_{i-1} & -s\alpha_{i-1} & 0 \\ 0 & s\alpha_{i-1} & c\alpha_{i-1} & 0 \\ 0 & 0 & 0 & 1 \end{bmatrix}, \tag{8}$$

is given by a rotation of α_{i-1} about x_{i-1}. Note that no translation appears in the matrix. That matrix makes the z_{i-1} parallel to the next, z_i-axis. The next matrix,

$$\begin{bmatrix} 1 & 0 & 0 & a_{i-1} \\ 0 & 1 & 0 & 0 \\ 0 & 0 & 1 & 0 \\ 0 & 0 & 0 & 1 \end{bmatrix}, \tag{9}$$

refers to the translation a_{i-1}. This translation is along the x_{i-1}-axis, defining the distance a_{i-1}. As z_{i-1} is parallel to z_i, the distance a_{i-1} is a mutual perpendicular to both the previous and the next z-axis. The rotation matrix,

$$\begin{bmatrix} c\theta_i & -s\theta_i & 0 & 0 \\ s\theta_i & c\theta_i & 0 & 0 \\ 0 & 0 & 1 & 0 \\ 0 & 0 & 0 & 1 \end{bmatrix}, \tag{10}$$

results of rotation about z_i by the joint angle θ_i . This rotation generates a frame parallel to $\{i\}$ at a distance d_i (offset) from the joint i. The 4th matrix,

$$\begin{bmatrix} 1 & 0 & 0 & 0 \\ 0 & 1 & 0 & 0 \\ 0 & 0 & 1 & d_i \\ 0 & 0 & 0 & 1 \end{bmatrix}, \tag{11}$$

translates the frame by d_i completing the transform from frame $\{i\}$ to $\{i-1\}$. Sometimes, it is more convenient to write the kinematic differential equations in the frame $\{i\}$. In this case, we have to transpose the rotation matrices so as to have the equations written in frame $\{i\}$ but in coordinates of $\{i-1\}$.

For space robots, we must compute the kinematics not only of the manipulator but also for the whole spacecraft (rotational and translational kinematics) [3, 4]. To do this, it is necessary to define three more frames:

- the Earth-centered inertial (ECI) frame $\bigoplus XYZ$, defined at the center of the Earth (the symbol $\bigoplus$ means Earth)
- the local vertical, local horizontal frame (LVLH) denoted as $Cx_l y_l z_l$. C is the center of mass of the space robot. The subscript l is a short for LVLH;
- and a spacecraft fixed system of reference, defined in the center of mass of the space robots [7] and denoted by $Cx_p y_p z_p$.

Figure 3 illustrates these new frames. The $Cx_l y_l z_l$ is fixed in the orbit. So, it follows with the orbit synchronized with the orbital velocity and with z_l pointing toward the local vertical (Nadir direction). In the literature, this frame is also named *RPY* (roll, pitch, and yaw). Roll is the rotation angle about the orbital velocity direction (x-direction), pitch is the rotation angle about the negative orbit normal (y-axis), and yaw is the rotation angle about the local vertical (z-axis). The roll, pitch, and yaw angles yield the space robot attitude with respect to the $Cx_l y_l z_l$ frame. Both the $Cx_l y_l z_l$ and the $Cx_p y_p z_p$ frames coincide with each other in a nominal attitude specification, that is, the space robot is pointing toward the local vertical, and the spacecraft attitude is zero. As the $Cx_p y_p z_p$ frame is fixed in the spacecraft, the attitude is given by the angles between $Cx_l y_l z_l$ and $Cx_p y_p z_p$. Other conventions may be used for the definition of these frames. The reference systems $\bigoplus XYZ$, $Cx_l y_l z_l$, and $Cx_p y_p z_p$ are included in the formulation to allow for describing the rotational and translational kinematics in frame $Cx_p y_p z_p$ or in the frame $Cx_l y_l z_l$. In general, the equations of dynamics are written in the body fixed frame, $Cx_p y_p z_p$. We can write the kinematics and/or the kinematics' differential equations in any of those three frames by using transformation matrices. The {ECI} allows writing the translational kinematic expressions in terms of the orbit parameters [7]. If the coupling between the orbit and attitude motion is negligible (in general, it is), we can write the rotational kinematics in the $Cx_p y_p z_p$ or $Cx_l y_l z_l$ frames, including the orbital velocity only. The practice is to use the Euler angles to describe the angular orientation between both frames (the attitude of the spacecraft with respect the $Cx_ly_lz_l$ frame). According to the Euler Theorem, we can do this through three sequences of rotations from one frame to another frame. **Figure 3** shows both frames in a nominal configuration in which the attitude angles are zero and both frames are coincident. As the spacecraft moves, the fixed frames also move. **Figure 4** illustrates the rotation between those frames. So, we can derive the kinematic expressions by using the spacecraft fixed frame, $x_py_pz_p$.

Let us consider that the spacecraft fixed frame is misaligned from the $Cx_ly_lz_l$ by the angles as shown in **Figure 4**. **Figure 4a** illustrates the space robot in-orbit configuration with attitude different from zero. Note that in this figure, both $Cx_ly_lz_l$ and $Cx_py_pz_p$ are misaligned. **Figure 4b** illustrates the sequence of rotations from $Cx_ly_lz_l$ to $Cx_py_pz_p$.

The sequence of three rotations is $\varphi \longleftarrow \theta \longleftarrow \psi$ [7]. The arrows indicate the direction of the sequence as: first, a rotation about z_l by the angle ψ generating an auxiliary frame $Cx'_py'_pz'_p$. Then, a second rotation is done about y'_l by the angle θ generating the auxiliary frame $Cx''_py''_pz''_p$. Finally, the third rotation is done about x'' to coincide with $Cx_py_pz_p$. There are 12 sets of Euler angle sequences of rotations to describe the rotational kinematics. However, the sequence of rotations $\varphi \longleftarrow \theta \longleftarrow \psi$ is one the most used transformations for satellites' rotational kinematics description. This transform provides the relationship between both $Cx_ly_lz_l$ and $Cx_py_pz_p$ frames. The notation of the sequence used here is the same as $\varphi \longrightarrow \theta \longrightarrow \psi$. However, the first notation emphasizes the resulting structure of matrix multiplication when implementing the transformation from $Cx_ly_lz_l$ to $Cx_py_pz_p$ frames as we are going to show next.

First rotation, about z_l

$$\begin{Bmatrix} x'_p \\ y'_p \\ z'_p \end{Bmatrix} = \begin{bmatrix} c\psi & s\psi & 0 \\ -s\psi & c\psi & 0 \\ 0 & 0 & 1 \end{bmatrix} \begin{Bmatrix} x_l \\ y_l \\ z_l \end{Bmatrix} \quad (12)$$

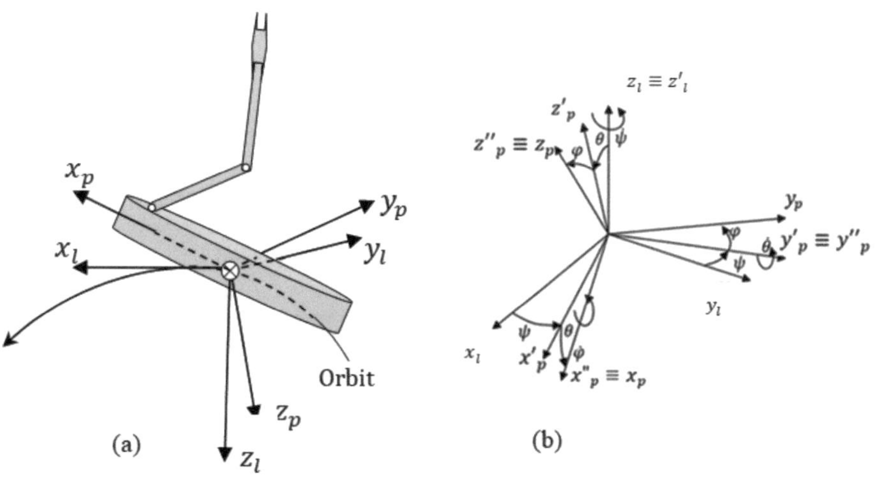

Figure 4.
(a) Satellite on-orbit configuration with $Cx_py_pz_p$ misaligned from $Cx_ly_lz_l$ and (b) the Euler sequence of rotations $\varphi \longrightarrow \theta \longrightarrow \psi$.

Second rotation, about $y'_p \equiv y''_p$

$$\begin{Bmatrix} x''_p \\ y''_p \\ z''_p \end{Bmatrix} = \begin{bmatrix} c\theta & 0 & -s\theta \\ 0 & 1 & 0 \\ s\theta & 0 & c\theta \end{bmatrix} \begin{Bmatrix} x'_p \\ y'_p \\ z'_p \end{Bmatrix} \tag{13}$$

Third rotation, about $x''_p \equiv x_p$

$$\begin{Bmatrix} x_p \\ y_p \\ z_p \end{Bmatrix} = \begin{bmatrix} 1 & 0 & 0 \\ 0 & c\varphi & s\varphi \\ 0 & -s\varphi & c\varphi \end{bmatrix} \begin{Bmatrix} x''_p \\ y''_p \\ z''_p \end{Bmatrix} \tag{14}$$

By substituting Eq. (13) into Eq. (14) and substituting Eq. (12) into the result, we have:

$$\begin{Bmatrix} x_p \\ y_p \\ z_p \end{Bmatrix} = \begin{bmatrix} 1 & 0 & 0 \\ 0 & c\varphi & s\varphi \\ 0 & -s\varphi & c\varphi \end{bmatrix} \begin{bmatrix} c\theta & 0 & -s\theta \\ 0 & 1 & 0 \\ s\theta & 0 & c\theta \end{bmatrix} \begin{bmatrix} c\psi & s\psi & 0 \\ -s\psi & c\psi & 0 \\ 0 & 0 & 1 \end{bmatrix} \begin{Bmatrix} x_l \\ y_l \\ z_l \end{Bmatrix} \tag{15}$$

So, the structure is really like $\varphi \longleftarrow \theta \longleftarrow \psi$.

The final matrix allows writing the equations in the spacecraft fixed frame $Cx_py_pz_p$ in coordinates of $Cx_ly_lz_l$ as:

$$\begin{Bmatrix} x_p \\ y_p \\ z_p \end{Bmatrix} = \begin{bmatrix} c\theta c\psi & c\theta s\psi & -s\theta \\ s\varphi s\theta c\psi - c\varphi s\psi & s\varphi s\theta s\psi + c\varphi c\psi & s\varphi c\theta \\ c\varphi s\theta c\psi + s\varphi s\psi & c\varphi s\theta s\psi - s\varphi c\psi & c\varphi c\theta \end{bmatrix} \begin{Bmatrix} x_l \\ y_l \\ z_l \end{Bmatrix} = {}^p_l\boldsymbol{R} \tag{16}$$

By transposing this matrix, we can write the kinematics' equation in the LVLH frame:

$$\begin{Bmatrix} \mathrm{x_l} \\ \mathrm{y_l} \\ \mathrm{z_l} \end{Bmatrix} = \begin{bmatrix} \mathrm{c\theta c\psi} & \mathrm{c\theta s\psi} & -s\theta \\ \mathrm{s\varphi s\theta c\psi - c\varphi s\psi} & \mathrm{s\varphi s\theta s\psi + c\varphi c\psi} & \mathrm{s\varphi c\theta} \\ \mathrm{c\varphi s\theta c\psi + s\varphi s\psi} & \mathrm{c\varphi s\theta s\psi - s\varphi c\psi} & \mathrm{c\varphi c\theta} \end{bmatrix}^{\mathrm{T}} \begin{Bmatrix} \mathrm{x_p} \\ \mathrm{y_p} \\ \mathrm{z_p} \end{Bmatrix} = {}^l_p\mathrm{R}\{\mathrm{p}\} \tag{17}$$

The relationships from $\{\mathrm{n}\}$ to $\{0\}$ can be written as:

$${}^0_nT = {}^0_1T\,{}^1_2T\,{}^2_3T \ldots {}^{n-1}_{\;\;n}T \tag{18}$$

To write the kinematics of the spacecraft-like robot manipulator in the LVLH system of reference, we have to concatenate all the matrices. By using Eqs. (17) and (18) and Eq. (2), we can establish the relationships between i the coordinate frames as:

$${}^l_iT = {}^l_pT\,{}^p_0T\,{}^{i-1}_{\;\;i}T \tag{19}$$

where

$${}^l_pT = \begin{bmatrix} {}^l_p\boldsymbol{R}_{3\times 1} & \boldsymbol{0}_{3\times 1} \\ \boldsymbol{0}_{1\times 3} & 1 \end{bmatrix} = \begin{bmatrix} c\theta c\psi & s\varphi s\theta c\psi - c\varphi s\psi & c\varphi s\theta c\psi + s\varphi s\psi & r_{cx} \\ c\theta s\psi & s\varphi s\theta s\psi + c\varphi c\psi & s\varphi c\theta & r_{cy} \\ -s\theta & s\varphi c\theta & c\varphi c\theta & r_{cz} \\ 0 & 0 & 0 & 1 \end{bmatrix};$$

DOI: http://dx.doi.org/10.5772/intechopen.85636

$${}_{0}^{p}\mathrm{T} = \begin{bmatrix} -1 & 0 & 0 \\ 0 & 0 & 1 \\ 0 & 1 & 0 \end{bmatrix}$$ —this transform relates $\mathrm{Ox_O y_O z_O}$ with $\mathrm{Cx_p y_p z_p}$ (see **Figure 3**)

${}_{i}^{i-1}T$ is given by Eq. (2)

$$\mathrm{i} = 1, 2, \ldots, \mathrm{n}$$

The vector $\{r_c\} = \{r_{cx}\ r_{cy}\ r_{cz}\}^T$ represents the translation from the center of mass (CM) to the robot base or the $Ox_O y_O z_O$ frame. This is a constant vector in this formulation. However, when the robot manipulator is in orbit operation, the links move, changing the location of the system CM, in other words, the CM location becomes a variable of the problem. By the time the space shuttle was in operation, this type of problem appeared. When it grasped significant massive target, the system CM motion affected the efficiency of the robot arm to put the target where it was planned. Another problem experienced by the space shuttle manipulator was the interaction between the attitude control system and the robot arm structural flexibility. The attitude control was in action to keep the shuttle stabilized while the manipulator was commanded to grasp a target. The structural flexibility of the long arms was excited by the attitude actuators and entered in elastic vibration mode causing problems to accomplish a safety grasping of the target. But this is another story and may be subject of research on another occasion.

Example 1. Consider the robot manipulator shown in **Figure 3** where the space robot is stabilized with $Cx_l y_l z_l$ and $Cy_p z_p$ frames. Find the kinematics relationship written in the $Ox_O y_O z_O$ frame. First, we built the D-H as shown in **Table 1** for the problem.

By inspection in **Figure 3**, we can see that all the z-axes are parallel and the joints are revolute. So, the parameters α and d are zeros. Following the D-H table:

$${}_{1}^{0}T = \begin{bmatrix} c\theta_1 & -s\theta_1 & 0 & 1 \\ s\theta_1 & c\theta_1 & 0 & 0 \\ 0 & 0 & 1 & 0 \\ 0 & 0 & 0 & 1 \end{bmatrix} \tag{20}$$

$${}_{2}^{1}T = \begin{bmatrix} c\theta_2 & -s\theta_2 & 0 & l_1 \\ s\theta_2 & c\theta_2 & 0 & 0 \\ 0 & 0 & 1 & 0 \\ 0 & 0 & 0 & 1 \end{bmatrix} \tag{21}$$

i	α_{i-1}	a_{i-1}	d_i	θ_i
1	0	0	0	θ_1
2	0	l_1	0	θ_2
3	0	l_2	0	θ_3

Table 1.
D-H parameters for Example 1.

$$
{}^{2}_{3}T = \begin{bmatrix} c\theta_3 & -s\theta_3 & 0 & l_2 \\ s\theta_3 & c\theta_3 & 0 & 0 \\ 0 & 0 & 1 & 0 \\ 0 & 0 & 0 & 1 \end{bmatrix} \tag{22}
$$

so

$$
{}^{0}_{3}T = {}^{0}_{1}T^{1}_{2}T^{2}_{3}T = \begin{bmatrix} c\theta_1 & -s\theta_1 & 0 & 1 \\ s\theta_1 & c\theta_1 & 0 & 0 \\ 0 & 0 & 1 & 0 \\ 0 & 0 & 0 & 1 \end{bmatrix} \begin{bmatrix} c\theta_2 & -s\theta_2 & 0 & l_1 \\ s\theta_2 & c\theta_2 & 0 & 0 \\ 0 & 0 & 1 & 0 \\ 0 & 0 & 0 & 1 \end{bmatrix} \begin{bmatrix} c\theta_3 & -s\theta_3 & 0 & l_2 \\ s\theta_3 & c\theta_3 & 0 & 0 \\ 0 & 0 & 1 & 0 \\ 0 & 0 & 0 & 1 \end{bmatrix} \tag{23}
$$

Compacting the result through the definition of c = *cos*, s = *sin*, and $i = \theta_i, i = 1, 2, 3$ we have

$$
{}^{0}_{3}T = \begin{bmatrix} c_3c_{12} - s_3s_{12} & -c_3s_{12} - s_3c_{12} & 0 & l_1c_1 + l_2c_{12} \\ c_3s_{12} + s_3c_{12} & c_3c_{12} - s_3s_{12} & 0 & l_1s_1 + l_2s_{12} \\ 0 & 0 & 1 & 0 \\ 0 & 0 & 0 & 1 \end{bmatrix} = \begin{bmatrix} c_{123} & -s_{123} & 0 & l_1c_1 + l_2c_{12} \\ s_{123} & c_{123} & 0 & l_1s_1 + l_2s_{12} \\ 0 & 0 & 1 & 0 \\ 0 & 0 & 0 & 1 \end{bmatrix} \tag{24}
$$

where

$$
c_{123} = \cos\left(\sum_{i=1}^{3} \theta_i\right) = c_3c_{12} - s_3s_{12} \tag{25}
$$

$$
s_{123} = \sin\left(\sum_{i=1}^{3} \theta_i\right) = c_3s_{12} + s_3c_{12} \tag{26}
$$

$$
\mathrm{i} = 1, 2, 3
$$

Note that we considered the spacecraft stable in attitude with the $Cx_py_pz_p$ aligned to the $Cx_ly_lz_l$ frame. But be aware that if the manipulator moves in attitude, the kinematics' differential equations of the spacecraft shall account for the manipulator links of kinematics' differential equations. In particular, if the manipulator is to grasp an on-orbit target, the attitude of the manipulator-like spacecraft must be synchronized to that of the target so that the relative attitude of both spacecraft and target is zero. As the dynamic analysis is out of the scope of this chapter, we will assume that the robot arm kinematics' equations differ from ground robots only by $Cx_ly_lz_l$ and $x_py_pz_p$ frames. For orbital robots these frames are used to describe the kinematics differential equations related to orbit and attitude motions.

Example 2. Consider **Figure 5** which illustrates a revolute-prismatic-revolute (RPR) manipulator. In this case, we have a rotation (θ_1), a translational variable (d_2), and another rotation (θ_2) as illustrated in **Table 2**. The convention is that those degrees-of-freedom occur along z-axis.

For this case, we have one rotation of θ_1, translation of d_2, and a rotation of θ_3.
Following the D-H convention above, we should make a few comments. First, we have a rotation on z-axis. Then, we use the twist angle rotation to make a

DOI: http://dx.doi.org/10.5772/intechopen.85636

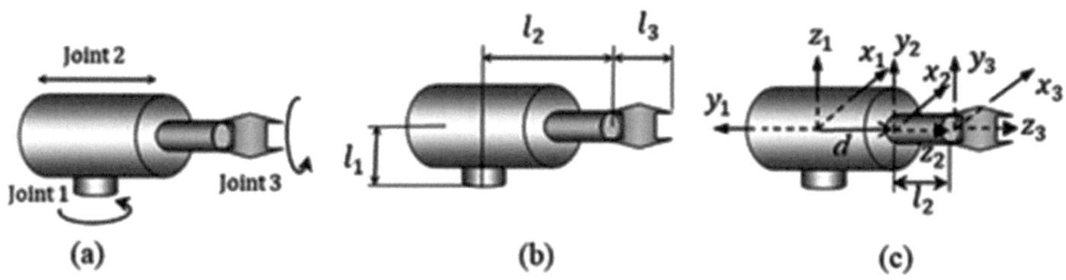

Figure 5.
(a) RPR manipulator showing the joints, (b) D-H parameters, and (c) frames.

new z-axis for the joint variable d_2. The sequence of rotations is shown in Eqs. (18)–(20).

$$ {}^{0}_{1}T = \begin{bmatrix} c\theta_1 & -s\theta_1 & 0 \\ s\theta_1 & c\theta_1 & 0 \\ 0 & 0 & a1 \end{bmatrix} \tag{27} $$

$$ {}^{1}_{2}T = \begin{bmatrix} 1 & 0 & 0 \\ 0 & c\alpha_1 & -s\alpha_1 \\ 0 & c\alpha_1 & c\alpha_1 \end{bmatrix} \tag{28} $$

$$ {}^{2}_{3}T = \begin{bmatrix} c\theta_3 & -s\theta_3 & 0 \\ s\theta_3 & c\theta_3 & 0 \\ 0 & 0 & 1 \end{bmatrix} \tag{29} $$

To account for the translation d_2, we can use the homogeneous transformation as:

$$ \begin{bmatrix} c\theta_1 & -s\theta_1 & 0 & 0 \\ s\theta_1 & c\theta_1 & 0 & 0 \\ 0 & 0 & 1 & 0 \\ 0 & 0 & 0 & 1 \end{bmatrix} \tag{30} $$

$$ \begin{bmatrix} 1 & 0 & 0 & 0 \\ 0 & c\alpha_1 & -s\alpha_1 & 0 \\ 0 & s\alpha_1 & c\alpha_1 & 0 \\ 0 & 0 & 0 & 1 \end{bmatrix} \tag{31} $$

$$ \begin{bmatrix} 1 & 0 & 0 & 0 \\ 0 & 1 & 0 & 0 \\ 0 & 0 & 1 & d_2 \\ 0 & 0 & 0 & 0 \end{bmatrix} (\alpha_1 = 90^o) \tag{32} $$

$$ \begin{bmatrix} c\theta_3 & -s\theta_3 & 0 & 0 \\ s\theta_3 & c\theta_3 & 0 & 0 \\ 0 & 0 & 1 & l_2 \\ 0 & 0 & 0 & 1 \end{bmatrix} \tag{33} $$

Note that this notation of homogeneous matrix allows including the two translations of this problem. Also, note that the offset translation is along z. The transform matrix ($\alpha_2 = 90^o$) yields:

$$\begin{bmatrix} 1 & 0 & 0 & 0 \\ 0 & c\alpha_1 & -s\alpha_1 & 0 \\ 0 & s\alpha_1 & c\alpha_1 & 0 \\ 0 & 0 & 0 & 1 \end{bmatrix} \begin{bmatrix} 1 & 0 & 0 & 0 \\ 0 & 1 & 0 & 0 \\ 0 & 0 & 1 & d_2 \\ 0 & 0 & 0 & 0 \end{bmatrix} = \begin{bmatrix} 1 & 0 & 0 & 0 \\ 0 & 0 & -1 & 0 \\ 0 & 1 & 0 & 0 \\ 0 & 0 & 0 & 1 \end{bmatrix} \begin{bmatrix} 1 & 0 & 0 & 0 \\ 0 & 1 & 0 & 0 \\ 0 & 0 & 1 & d_2 \\ 0 & 0 & 0 & 0 \end{bmatrix} = \begin{bmatrix} 1 & 0 & 0 & 0 \\ 0 & 0 & -1 & -d_2 \\ 0 & 1 & 0 & 0 \\ 0 & 0 & 0 & 1 \end{bmatrix} \tag{34}$$

This result is a transformation, ${}^{1}_{2}T^{\,2}$, that is, the transform from frame {2} to frame {1}. Combining all the transformations, we can write from frame {3} to frame {0} as:

$${}^{0}_{3}T = {}^{0}_{1}T\,{}^{1}_{2}T\,{}^{2}_{3}T = \begin{bmatrix} c\theta_1 & -s\theta_2 & 0 & 0 \\ s\theta_2 & c\theta_1 & 0 & 0 \\ 0 & 0 & 1 & 0 \\ 0 & 0 & 0 & 1 \end{bmatrix} \begin{bmatrix} 1 & 0 & 0 & 0 \\ 0 & 0 & -1 & -d_2 \\ 0 & 1 & 0 & 0 \\ 0 & 0 & 0 & 1 \end{bmatrix} \begin{bmatrix} c\theta_3 & -s\theta_3 & 0 & 0 \\ s\theta_3 & c\theta_3 & 0 & 0 \\ 0 & 0 & 1 & 0 \\ 0 & 0 & 0 & 1 \end{bmatrix} = \begin{bmatrix} c\theta_1 c\theta_3 & -c\theta_1 s\theta_3 & s\theta_3 & s\theta_1(l_2 + d_2) \\ s\theta_1 c\theta_3 & -s\theta_1 s\theta_3 & -c\theta_1 & -c\theta_1(l_2 + d_2) \\ s\theta_3 & c\theta_3 & 0 & 0 \\ 0 & 0 & 0 & 1 \end{bmatrix} \tag{35}$$

We can also find the transpose of this transformation to have the transformation:

$${}^{3}_{0}T = {}^{3}_{2}T\,{}^{2}_{1}T\,{}^{1}_{0}T \begin{bmatrix} c\theta_3 & s\theta_3 & 0 & 0 \\ -s\theta_3 & c\theta_3 & 0 & 0 \\ 0 & 0 & 1 & 0 \\ 0 & 0 & 0 & 1 \end{bmatrix} \begin{bmatrix} 1 & 0 & 0 & 0 \\ 0 & 0 & 1 & -d_2 \\ 0 & -1 & 0 & 0 \\ 0 & 0 & 0 & 1 \end{bmatrix} \begin{bmatrix} c\theta_1 & s\theta_1 & 0 & 0 \\ -s\theta_1 & c\theta_1 & 0 & 0 \\ 0 & 0 & 1 & 0 \\ 0 & 0 & 0 & 1 \end{bmatrix} = \begin{bmatrix} c\theta_1 c\theta_3 & s\theta_1 c\theta_3 & s\theta_3 & 0 \\ -c\theta_1 s\theta_3 & -s\theta_1 s\theta_3 & c\theta_3 & 0 \\ s\theta_1 & -c\theta_1 & 0 & -(l_2 + d_2) \\ 0 & 0 & 0 & 1 \end{bmatrix} \tag{36}$$

3. Inverse kinematics

Inverse kinematics consists of finding the joint variables given the robot manipulator end-effector position and orientation with respect to the user workstation. The workstation means a known location and reference system. It could be the platform of the robotic system. In this chapter, the platform reference system is the frame $Cx_p y_p z_p$ represented by {P}. In general, such frame is defined in the center of mass of the space robotic system. We know the location of that system of reference. Also, we know the location of the $Cx_ly_lz_l$ frame because we know the space robot orbit. In the previous section, we considered the relationship between frames to write the equations in the {0}. In the inverse problem, we want to find joint variables given the specified position and orientation of the tool relative to {0} or {p}. We assume that the robot attitude and orbit control subsystem keeps the space robot stabilized while it works. This is to simplify the formulation. This assumption is reasonable since this is exactly what must be done when the space robot is in operation. We have already presented and discussed the several frames defined to study the robot kinematics. We will refer to just one more, the frame defined on the tool, {T}.

There are some aspects of inverse kinematics that make it more complex than forward kinematics [5, 6]. We will briefly report them, but we will not go into too many details. In this chapter, our focus is on the solution methods for inverse kinematics problem.

One of the main aspects of the inverse kinematics is solvability because we have to face the problem of existence of solutions and of multiple solutions. Also, this problem brings to light the concept of manipulator workspace which is the volume of space that the end-effector of the manipulator can reach. For space robot, this problem is critical because if the object to be manipulated is beyond space robot workspace, the space mission may not accomplish goals like grasping a target, docking to another spacecraft, put a piece of structure in right location during assembling space structures, and so on. For a solution to exist, the point specified to be reached by the end-effector must be inside the manipulator workspace. Another problem related to the workspace of robots is to plan a trajectory [9] preventing collisions with objects, another robot, or astronaut inside the workspace. There are two useful definitions related to workspace. One is the dextrous workspace [5] and the other is the reachable workspace. The first refers to the space volume that the end-effector reaches with all orientation. The second is that volume of space that the robot can reach in at least one orientation. If the position and orientation specified for the robot task is inside the workspace, then at least one solution exists. Workspace also depends on the tool-frame transformation, since it is frequently the tool-tip that is discussed when we discuss reachable points in space. Details about the solvability problem for inverse kinematics may be seen in [5].

3.1 The inverse kinematics solution methods

In solving inverse problems, the first thing we must consider is that there are no general algorithms [5, 8, 9] that may be employed to solve manipulator kinematics. We state that a manipulator is solvable when the joint variables can be determined by an algorithm for a given position and orientation of the tool frame, {T}.

The solutions may be classified as close solution and numerical solutions. Numerical solutions are in general slower than close solutions and would require another chapter for discussion. We will concentrate in the close solutions methods. These methods can be subdivided into algebraic and geometric solutions. This classification is somehow hazy since each solution method uses both algebraic and geometric kinematic formulations. In presenting and discussing the methods, we point out that according to the concept of solution we stated here, the revolute and prismatic manipulators having six degrees of freedom (DOF) in series chain are solvable.

3.2 Algebraic solution

Consider the robot manipulator mounted on the spacecraft platform and the on-orbit configurations in the nominal attitude as shown in **Figure 3**. Let us apply the algebraic solution. Note that the frame defined in the last link is the frame for n = 3. So, the transform from frame {0} to frame {N} is given by Eq. (24) repeated here for didactic reasons.

$$ {}^{O}_{N}T = \begin{bmatrix} c_{123} & -s_{123} & 0 & l_1c_1 + l_2c_{12} \\ s_{123} & c_{123} & 0 & l_1s_1 + l_2s_{12} \\ 0 & 0 & 1 & 0 \\ 0 & 0 & 0 & 1 \end{bmatrix} \tag{37} $$

Now consider that for inverse kinematics, the specified position of the end-effector is given:

$$ {}^{O}_{N}T = \begin{bmatrix} c\varphi & -s\varphi & 0 & x \\ s\varphi & c\varphi & 0 & y \\ 0 & 0 & 1 & 0 \\ 0 & 0 & 0 & 1 \end{bmatrix} \tag{38} $$

Note that this specified orientation and position are the angle φ and the Cartesian position (x, y, 0) of the tool. As Eq. (37) must be equal to Eq. (38), the problem can be solved by comparing the elements of the matrices. The joint angles and the parameters l_i, $i = 1, 2$ are unknown. The specified orientation given by φ and the components of the vector $\{x \quad y \quad z\}^T$, in the planar case, x and y are known and represent the tool position with respect to {0}. By equating a_{11} element of both matrices we have:

$$ c\varphi = c_{123} = \cos\left(\sum_{i=1}^{3}\theta_i\right) \tag{39} $$

$$ s\varphi = s_{123} = \cos\left(\sum_{i=1}^{3}\theta_i\right) \tag{40} $$

$$ x = l_1c_1 + l_2c_{12} \tag{41} $$

$$ y = l_1s_1 + l_2s_{12} \tag{42} $$

where

$$ c_{12} = c_1c_2 - s_1s_2 $$

$$ s_{12} = c_1s_2 + s_1c_2 $$

The algebraic solution requires the solution of Eq. (39) to Eq. (42). We can compute the square of Eqs. (41) and (42) to obtain:

$$ x^2 + y^2 = {l_1}^2 + {l_2}^2 + 2l_1l_2c_2 \tag{43} $$

Solving Eq. (43) by c_2, we compute c_2 as:

$$ c_2 = \frac{x^2 + y^2 - \left({l_1}^2 + {l_2}^2\right)}{2l_1l_2} \tag{44} $$

Important information can be extracted from Eq. (44). We know that the cosine function must be between −1 and + 1. By checking such constraint, we obtain information whether or not the solution exists. In case a solution does not exist, the meaning is that the point is not inside the robot manipulator workspace.

Let us assume that the orientation and position specified in inside the workspace. In this case, we can compute s_2 as:

$$s_2 = \pm\sqrt{1 - c_2{}^2} \tag{45}$$

The joint angle can then be computed by:

$$\theta_2 = \text{Atan2}\left(\frac{s_2}{c_2}\right) \tag{46}$$

$\text{Atan2}\left(\frac{s_2}{c_2}\right)$ computes $\tan^{-1}\left(\frac{s_2}{c_2}\right)$, but uses the signs of both s_2 and c_2 to identify the quadrant in which the resulting angle lies.

Note that there are signs to be considered in Eq. (45). The signs are related to multiple solutions.

Once we have obtained θ_2, we can obtain θ_1 by manipulating Eqs. (41) and (42). To solve the problem, we should define:

$$k_1 = l_1 + l_2 c_2 \tag{47}$$

$$k_2 = s_2 \tag{48}$$

and use the trigonometric identities

$$c_{12} = c_1 c_2 - s_1 s_2 \tag{49}$$

$$s_{12} = c_1 s_2 + s_1 c_2 \tag{50}$$

to rewrite Eqs. (41) and (42) as:

$$x = k_1 c_1 - k_2 s_1 \tag{51}$$

$$x = k_1 s_1 + k_2 c_1 \tag{52}$$

Next, we make a change of variable by writing k_1 and k_2 in the form:

$$r = +\sqrt{k_1{}^2 + k_2{}^2} \tag{53}$$

and

$$\gamma = \text{Atan2}(k_1, k_2) \tag{54}$$

Using these new variables, we can write Eqs. (51) and (52) as:

$$\frac{x}{r} = c\gamma c\theta_1 - s\gamma c s\theta_1 = c(\gamma + \theta_1) \tag{55}$$

$$\frac{y}{r} = c\gamma s\theta_1 + s\gamma\theta_1 = s(\gamma + \theta_1) \tag{56}$$

from which

$$\gamma + \theta_1 = \text{Atan2}\left(\frac{y}{r}, \frac{x}{r}\right) = \text{Atan2}(y, x) \tag{57}$$

Then, using the definition of γ, we obtain the expression for θ_1 as:

$$\theta_1 = \text{Atan2}(y, x) - \text{Atan2}(k_1, k_2) \tag{58}$$

As we have θ_1 and θ_2, we can find θ_3 by using Eqs. (39) and (40)

$$\sum_{i=1}^{3} \theta_i = \text{Atan2}(s\varphi, c\varphi) = \varphi = \theta_1 + \theta_2 + \theta_3 \tag{59}$$

Once have θ_1 and θ_2, we obtain θ_3 from Eq. (59).

Transcendental equations commonly arise when applying the algebraic approach. The greater the number of degrees of freedom of the robotic manipulator, the more laborious is the solution of the problem.

3.3 Geometric solution

By using this approach, we simply work the geometry of the arm. For the same problem, we may have the position of the joint associated to the frame {N} with respect to the origin of frame {O} by using geometry. Frame {N} refers to the frame attached to the last link of the manipulator. Consider **Figure 6** showing the triangle formed by the geometrical configuration of the sides of the links and the line from the origin of the frame {O} to the origin of frame {N}. Applying the law of cosines, we can write:

$$x^2 + y^2 = {l_1}^2 + {l_2}^2 - 2l_1 l_2 \cos(180 - \theta_2) = {l_1}^2 + {l_2}^2 + 2l_1 l_2 c_2 \tag{60}$$

$$c_2 = \frac{x^2 + y^2 - ({l_1}^2 + {l_2}^2)}{2l_1 l_2} \tag{61}$$

Again, this solution must be between 1 and $-$ 1 in order to have a valid solution. Assuming that the solution exists, then:

$$s_2 = \sqrt{1 - \left[\frac{x^2 + y^2 - ({l_1}^2 + {l_2}^2)}{2l_1 l_2}\right]^2} \tag{62}$$

Finally, we can write another solution for θ_2 in terms of the tangent since we have both sin and cosine of the angle θ_2

$$\theta_2 = \text{Atan2}\left\{\pm\sqrt{1 - \left[\frac{x^2 + y^2 - ({l_1}^2 + {\ell_2}^2)}{2l_1 l_2}\right]^2}, \frac{x^2 + y^2 - ({l_1}^2 + {l_2}^2)}{2l_1 l_2}\right\} \tag{63}$$

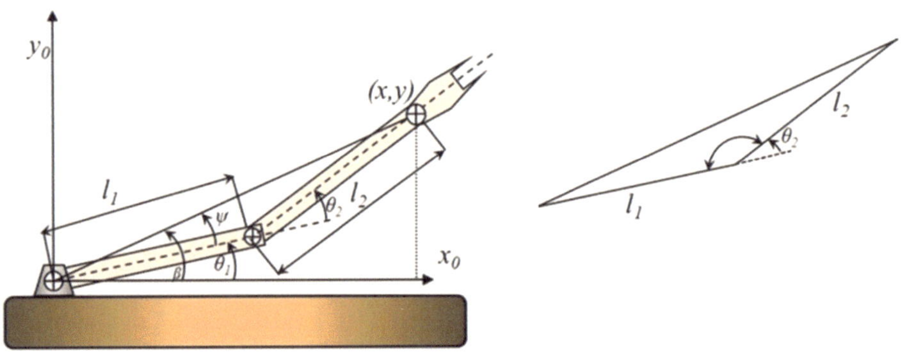

Figure 6.
Space robot manipulator—geometric solutions for the inverse kinematics.

DOI: http://dx.doi.org/10.5772/intechopen.85636

Finally, we can write other solution for θ_2 in terms of the tangent since we have both sin of and cosine of the angle θ_2

Now Let us compute a solution for θ_1

$$\beta = \text{Atan2}(y, x) \tag{64}$$

For ψ, we use the law of cosines to obtain:

$$c\psi = \frac{x^2 + y^2 + {l_1}^2 - {l_2}^2}{2 l_1 l_2} \tag{65}$$

Geometry yields then

$$\theta_1 = \beta \pm \psi \tag{66}$$

where $\beta = \text{Atan2}(y, x)$.

ψ is obtained by using the law of cosines as ${\ell_2}^2 = x^2 + y^2 + {l_1}^2 - 2\sqrt{x^2 + y^2} l_1 cos\psi$

By solving this equation by $cos\psi$, we obtain Eq. (65). Finally, we use Eqs. (64) and (65), to obtain θ_1 as in Eq. (66).

Now, we can compute θ_3 from:

$$\sum_{i=1}^{3} \theta_i = \theta_1 + \theta_2 + \theta_3 = \varphi \tag{67}$$

Solving Eq. (67) for θ_3, we complete the solution. Note that φ is the specified orientation and we had already computed θ_1 and θ_2. Thus, the problem is solved.

3.4 Algebraic solution method involving working matrix algebra

Let us consider a six-DOF revolute manipulator. The transformation from frames {O} to the {N} (N = 6) is

$$ {}^0_6T = \begin{bmatrix} a_{11} & a_{12} & a_{13} & p_x \\ a_{21} & a_{22} & a_{23} & p_y \\ a_{31} & a_{32} & a_{33} & p_z \\ 0 & 0 & 0 & 0 \end{bmatrix} \tag{68}$$

We know that this final matrix represents the following successive transforms

$$ {}^0_6T = {}^0_1T\,{}^1_2T\,{}^2_3T\,{}^3_4T\,{}^4_5T\,{}^5_6T \tag{69}$$

Where, for the revolute manipulator

$$ {}^0_1T = \begin{bmatrix} c_1 & -s_1 & 0 & 0 \\ s_1 & c_1 & 0 & 0 \\ 0 & 0 & 1 & 0 \\ 0 & 0 & 0 & 1 \end{bmatrix} \tag{70}$$

By premultiplying the left side of Eq. (68) by the inverse of Eq. (70), we obtain:

$$\begin{bmatrix} c_1 & s_1 & 0 & 0 \\ -s_1 & c_1 & 0 & 0 \\ 0 & 0 & 1 & 0 \\ 0 & 0 & 0 & 1 \end{bmatrix} \begin{bmatrix} a_{11} & a_{12} & a_{13} & p_x \\ a_{21} & a_{22} & a_{23} & p_y \\ a_{31} & a_{32} & a_{33} & p_z \\ 0 & 0 & 0 & 0 \end{bmatrix} = {}^1_2T\,{}^2_3T\,{}^3_4T\,{}^4_5T\,{}^5_6T \tag{71}$$

By equating both sides of Eq. (71), we start identifying solutions by comparing the elements of matrices from both sides. Then, we continue by multiplying both sides, by the inverse of the ${}^1_2T$ and repeat the process to identify more solutions to obtain:

$$\begin{bmatrix} c_2 & s_2 & 0 & -l_1c_2 \\ -s_2 & c_2 & 0 & l_1s_2 \\ 0 & 0 & 1 & 0 \\ 0 & 0 & 0 & 1 \end{bmatrix} \begin{bmatrix} c_1 & -s_1 & 0 & 0 \\ s_1 & c_1 & 0 & 0 \\ 0 & 0 & 1 & 0 \\ 0 & 0 & 0 & 1 \end{bmatrix} \begin{bmatrix} a_{11} & a_{12} & a_{13} & p_x \\ a_{21} & a_{22} & a_{23} & p_y \\ a_{31} & a_{32} & a_{33} & p_z \\ 0 & 0 & 0 & 0 \end{bmatrix} = {}^2_3T\,{}^3_4T\,{}^4_5T\,{}^5_6T \tag{72}$$

By implementing this process, we can solve the inverse kinematics for the six-DOF revolute joints with all the z-axis parallels.

Consider **Table 2** and the associated transformation matrix Eq. (36) for the RPR manipulator of **Figure 5**. Let us solve the inverse kinematics using the above method of matrix algebra. Al the matrix manipulations were computed by using the MATLAB® symbolic manipulator. The specified orientation and position is given by Eq. (68), by premultiplying the right side of this equation by inverse of Eq. (30), we obtain:

$$\begin{bmatrix} c_1 & s_1 & 0 & 0 \\ -s_1 & c_1 & 0 & 0 \\ 0 & 0 & 1 & 0 \\ 0 & 0 & 0 & 1 \end{bmatrix} \begin{bmatrix} a_{11} & a_{12} & a_{13} & p_x \\ a_{21} & a_{22} & a_{23} & p_y \\ a_{31} & a_{32} & a_{33} & p_z \\ 0 & 0 & 0 & 0 \end{bmatrix} = \begin{bmatrix} c\theta_3 & -s\theta_3 & 0 & 0 \\ 0 & 0 & -1 & -(l_2+d_2) \\ s\theta_3 & c\theta_3 & 0 & 0 \\ 0 & 0 & 0 & 1 \end{bmatrix}$$
$$= \begin{bmatrix} a_{11}c\theta_1 + a_{21}s\theta_1 & a_{12}c\theta_1 + a_{22}cs & a_{13}s\theta_1 & p_xc\theta_1 + p_ys\theta_1 \\ a_{21}c\theta_1 - a_{11}s\theta_1 & a_{22}c\theta_1 - a_{12}c\theta_1 & a_{23}c\theta_1 - a_{13}s\theta_1 & p_yc\theta_1 - p_xs\theta_1 \\ a_{31} & a_{32} & a_{33} & p_z \\ 0 & 0 & 0 & 1 \end{bmatrix} = {}^1_3T \tag{73}$$

Comparing the resulting matrices, on the left and right, for the elements (1,4) and (2,4), respectively, we obtain:

i	α_{i-1}	a_{i-1}	d_i	θ_i
1	0	0	0	θ_1
2	90	0	d_2	0
3	0	0	l_2	θ_3

Table 2.
The D-H parameters and joint variables.

$$\theta_1 = -\text{Atan2}\left(p_x, p_y\right) \tag{74}$$

$$d_2 = p_x s\theta_1 - p_y c\theta_1 - l_2 \tag{75}$$

Comparing elements from both sides yields

$$p_z = 0 \tag{76}$$

Premultiplying the result of left side again by the inverse of ${}^1_2T$, we obtain:

$$\begin{bmatrix} 1 & 0 & 0 & 0 \\ 0 & 0 & 1 & 0 \\ 0 & -1 & -d_3 & 0 \\ 0 & 0 & 0 & 1 \end{bmatrix} \begin{bmatrix} c_1 & s_1 & 0 & 0 \\ -s_1 & c_1 & 0 & 0 \\ 0 & 0 & 1 & 0 \\ 0 & 0 & 0 & 1 \end{bmatrix} \begin{bmatrix} a_{11} & a_{12} & a_{13} & p_x \\ a_{21} & a_{22} & a_{23} & p_y \\ a_{31} & a_{32} & a_{33} & p_z \\ 0 & 0 & 0 & 0 \end{bmatrix} = \begin{bmatrix} c\theta_3 & -s\theta_3 & 0 & 0 \\ s\theta_3 & c\theta_3 & 0 & 0 \\ s\theta_3 & c\theta_3 & 1 & l_2 \\ 0 & 0 & 0 & 1 \end{bmatrix} \tag{77}$$

The left side yields

$$\begin{bmatrix} a_{11}c\theta_1 + a_{21}s\theta_1 & a_{12}c\theta_1 + a_{22}s\theta_1 & a_{13}c\theta_1 + a_{23}s\theta_1 & p_x c\theta_1 + p_y s\theta_1 \\ a_{31} & a_{32} & a_{33} & p_z \\ a_{11}s\theta_1 - a_{21}c\theta_1 & a_{12}s\theta_1 - a_{22}c\theta_1 & a_{13}s\theta_1 - a_{23}c\theta_1 & p_x s\theta_1 - p_y c\theta_1 - d_2 \\ 0 & 0 & 0 & 1 \end{bmatrix} \tag{78}$$

By comparing the matrix elements (2,1) and (2,2) for both sides, we can compute θ_3 as:

$$a_{31} = s\theta_3 \text{ and } a_{32} = c\theta_3 \tag{79}$$

then, we obtain from $\tan\theta_3 = \frac{a_{31}}{a_{32}} \Longrightarrow \theta_3 = \text{Atan2}(a_{31}, a_{32})$

Of course, this example is simple. When we have a six-DOF manipulator, the algebraic manipulation to solve the inverse kinematics is hard, and the solution may require using transformation of variables. The MATLAB® software package has applications that easier the hard algebraic manipulation such as the symbolic manipulator and some specific functions for forward and inverse kinematics computation. MATLAB® functions include kinematics, trajectory generation, dynamics, and control. The toolbox also includes Simulink® models to describe the evolution of arm or mobile robot state over time for the sake of control. Regarding this chapter, functions like manipulating and converting between data types such as vectors, rotation matrices, homogeneous transformations, and twists are very important. Some important functions related to this chapter are:

- **angvec2r** that converts angle and vector orientation to a 3x3 rotation matrix;
- **angvec2tr** that converts angle and vector orientation to a 4x4 homogeneous transform;
- **syms** command that allows the definition of symbolic variables for rotation and homogeneous algebra. The symbolic manipulation allows include several commands to simplify results of matrix algebraic manipulation.
- **eul2tr (phi, theta, psi, options)** converts the Euler angles to a (4x4) homogeneous transformation matrix.

- **rpy2tr (roll, pitch, yaw, options)** command refers to the LVLH frame referred in this chapter and converts the roll-pitch-yaw angles to homogeneous transform.

There are several other commands/functions that are very useful for applications approached in this chapter and the reader is advised to see [10].

4. Conclusion

In this chapter, we have approached the forward and inverse kinematics for space robot manipulators via the D-H convention. In all cases presented here, the spacecraft nominal attitude was considered stabilized. Otherwise, it would be necessary to consider kinematics in terms of absolute position described in the LVLH or in the spacecraft fixed frame with variable attitude angles. Some examples are solved, and a related reference is provided. Al the algebraic matrix manipulations presented here were obtained by using the MATLAB® Symbolic tool box. Also, the chapter presented other features of the MATLAB® and Simulink® software related to the subject of the chapter. Despite its importance, differential kinematics is out of scope of this chapter and maybe the subject of future work.

Acknowledgements

Thanks to the Technological Institute of Aeronautics ITA/CTE and PG-EEC, National Institute for Space Research INPE/DMC, the Brazilian Space Agency (AEB), the Pontifical Catholic University of Sao Paulo PUC-SP, and FAPESP/PIPE —Project 00882-4/2017.

Author details

Ijar Milagre da Fonseca[1,2]*, Maurício Nacib Pontuschka[3]
and Glaydson Luiz Bertoze Lima[1]

1 Technological Institute of Aeronautics – DCTA/ITA, Sao Jose dos Campos, Brazil

2 National Institute for Space Research – DMC/INPE, Sao Jose dos Campos, Brazil

3 Pontifical Catholic University of Sao Paulo PUC-SP, Sao Paulo, Brazil

*Address all correspondence to: ijar@ita.br

References

[1] Fonseca I, Mint PM. The state-of-the-art in space robotics. Journal of Physics: Conference Series. 2015;**641**:7. DOI: 10.1088/1742-6596/641/1/012025

[2] Fonseca I, Pontuschka M, Saotome O, Bainum P, Lima G. The impact of the non-inertial base motion in the operations of robotic manipulators. In: 65th International Astronautical Congress. 2014

[3] Fonseca I, Goes L, Seito N, Duarte M, Mint OE. Attitude dynamics and control of a spacecraft like a robotic manipulator when implementing on-orbit servicing. Acta Astronautica. 2017; **137**:490-497. DOI: 10.1016/j. actaastro.2016.12.020

[4] Skaar SB, Ruoff CF. Teleoperation and Robotics in Space, Progress in Astronautics and Aeronautics. Vol. 61. Washington DC: American Institute of Aeronautics and Astronautics, Inc; 1994. pp. 257-258

[5] Craig J. Introduction to Robotics Mechanics and Control. 3rd ed. New York: Pearson; 2005. pp. 62-123. DOI: 10.13140/2.1.4486.1446

[6] Spong MW, Hutchinson S, Vidyasagar M. Robot Dynamics and Control. 2nd ed. ISBN: 978-0-471-64990-8. New York: John Wiley & Sons, Inc; 2004. pp. 57-89

[7] Wie B. Space Vehicle Dynamics and Control. 2nd ed. Reston, VA: American Institute of Aeronautics and Astronautics, Inc; 2012. pp. 397-381. DOI: 10.2514/4.860119

[8] Zohar I, Ailon A, Mint RR. Mobile robot characterized by dynamic and kinematic equations and actuator dynamics: Trajectory tracking and related application. Robotics and Autonomous Systems. 2011;**59**:343-353. DOI: 10.1016/j.robot.2010.12.001

[9] Song W, Hu GM. A fast inverse kinematics algorithm for joint animation. Procedia Engineering. 2011; **24**:350-354. DOI: 10.1016/j. proeng.2011.11.2655

[10] Corke P. Robotics, Vision & Control: Fundamental Algorithms in MATLAB. 2nd ed. Springer. 2011